W0007666

OPPORTUNITIES IN
INSTALLATION AND
REPAIR CAREERS

Mark Rowh

VGM Career Horizons
a division of *NTC Publishing Group*
Lincolnwood, Illinois USA

Cover Photo Credits

Clockwise from top left: DeVry Inc.; NTC Publishing Group; NTC Publishing Group; United Brotherhood of Carpenters and Joiners of America.

Library of Congress Cataloging-in-Publication Data
Rowh, Mark.
Opportunities in installation and repair careers / Mark Rowh.
p.   cm. — (VGM opportunities series)
Includes bibliographical references (p.   ).
ISBN 0-8442-4135-0 (hard) — ISBN 0-8442-4136-9 (soft)
1. Technologists—Vocational guidance.   2. Repairing trades—
Vocational guidance.   3. Installation of industrial equipment—
Vocational guidance.   I. Title.   II. Series.
TA157.R68   1994
602′.3—dc20   93-47510
                                                        CIP

Published by VGM Career Horizons, a division of NTC Publishing Group
4255 West Touhy Avenue
Lincolnwood (Chicago), Illinois 60646-1975, U.S.A.
© 1994 by NTC Publishing Group. All rights reserved.
No part of this book may be reproduced, stored in a retrieval system,
or transmitted in any form or by any means,
electronic, mechanical, photocopying, recording or otherwise,
without the prior permission of NTC Publishing Group.
Manufactured in the United States of America.

4 5 6 7 8 9 0 VP 9 8 7 6 5 4 3 2 1

ABOUT THE AUTHOR

Mark Rowh is a highly experienced educator in community college and vocational-technical education, as well as a widely published writer. He has held administrative positions at several postsecondary institutions, and currently serves on the administrative staff at New River Community College in Dublin, Virginia. Rowh holds a doctorate in vocational and technical education from Clemson University, and has worked closely with a variety of technical programs in his career as a professional educator.

Rowh is the author of ten other books, including several other volumes in the VGM Career Horizons series. In addition, his articles on educational and career topics have appeared in a wide range of magazines.

ACKNOWLEDGMENTS

The author greatly appreciates the cooperation of the following in the researching and writing of this book:

Career College Association
Corning Community College
Electronic Technicians Association
Greenville Technical College
Johnson County Community College
International Society of Certified Electronics
        Technicians
New River Community College
National Association of Trade and Technical Schools
National Automatic Merchandising Association
Northeast State Technical Community College
Ocean County College
Central Piedmont Community College
Penn Valley Community College
Pennsylvania College of Technology

Refrigeration Service Engineers Society
United States Telephone Association
U.S. Department of Labor
Virginia Department of Labor and Industry

FOREWORD

Installation and repair careers offer a world of opportunities to those seeking their first occupation as well as those interested in a job change. Flexibility, diversity, and challenge are all waiting in this dynamic field.

The installation and repair professional has the opportunity to apply his or her problem-solving and technical skills in a wide array of settings. Consider all the places in which valuable, complex equipment is essential to our lives—offices, hospitals, even our homes—and you can begin to imagine the demand for expert technicians to install and maintain these vital systems.

If you have the aptitude and the interest, the educational and career opportunities are available for you. By exploring careers in installation and repair you can find not only challenging, rewarding work but also good compensation and benefits. Inside you'll find all the information you need to explore this exciting career field.

The Editors of VGM Career Horizons

INTRODUCTION

Can you picture yourself making a career of installing, servicing, or repairing equipment or other items? Or do you know someone else who seems a good prospect for such a career? If so, this book is for you.

Choosing a career is among the most important decisions anyone makes. A career is more than just a way of earning a living. It is also part of the very definition of what makes one person alike or different from another. Perhaps most importantly, one's occupation represents a tremendous opportunity for self-fulfillment.

For people with the right aptitudes and skills, installation and repair fields hold great career potential. Such careers can be truly rewarding for people who enjoy solving problems, working with tools, and mastering the elements of electronics or other modern technologies.

Not only do technical careers of this type involve interesting work, they also provide good wages, benefits, and overall career potential. After all, society needs trained workers who can maintain and repair important equipment. Thus, for the right men and women, installation and career fields represent a promising career path.

The material provided in this book is designed to help people interested in this area explore their career options. It is the author's hope that the various details that follow will help the reader proceed positively in this direction.

This book is dedicated to
Jerry Dugger and Floyd Hogue,
two of the finest educators
I have had the privilege of knowing.

CONTENTS

CHAPTER 1

THE WORLD OF INSTALLATION AND REPAIR

As young people look to the future, the question of career choice looms large. What career area would be best for me? What talents or aptitudes do I have? How can I find an occupation that I will enjoy?

Similar questions also are faced by older adults who desire new career directions. Homemakers who have deferred career interests for family considerations, workers in "dead end" jobs, and other men and women looking for change often consider the question of which career to pursue, as well as how to get there.

Through personal interest, school performance, or other measures, many people find they have potential for working with equipment. Perhaps they enjoy tinkering with electronic devices. Maybe they have a flair for repairing motors or other equipment. Perhaps they are good with tools, and enjoy the challenge of analyzing and repairing problems when equipment at work or at home does not work properly.

1

In some cases, people with such aptitudes might be encouraged to become engineers. But engineering is a complex field, requiring at least four years of college, as well as a grasp of higher mathematics and scientific theory. At the other occupational extreme, they might become unskilled workers performing simple, unfulfilling job tasks. Or they might become production workers in industry, working on an assembly line or performing other repetitive tasks that, over the long haul, provide few career challenges.

A good compromise is offered by occupations where men and women function as technicians. Technicians perform tasks that are technical in nature but do not require the advanced theoretical work and depth of training needed by engineers. Technicians who specialize in various areas of installation and repair fall within this occupational category. They perform work that is important, challenging, and varied but that can be undertaken without a four-year college degree. In addition, they hold jobs in fields with a solid potential for earning good salaries and a continuing demand for trained workers.

THE NEED FOR INSTALLATION AND REPAIR

To what extent do you rely on mechanical or electronic devices as a part of daily life? Unless you live in a monastery, the answer is probably a great deal. A typical home, for instance, includes most or all of the following items: refrigerator, stove, furnace or heat pump, microwave oven, television, videocassette player, radio or stereo system, personal computer, and other items used for cooking, climate control, or entertainment. In addition, other devices such as lawnmowers and power tools are common.

The workplace features many more kinds of equipment. Telephones, fax machines, copiers, vending machines, alarm systems, construction equipment, large-scale computer systems, and more help workers perform their jobs.

Many of these items require the attention of technicians who are qualified to install, service, or repair them. As a result, hundreds of thousands of women and men make careers of providing the technical work needed in this area.

TYPES OF CAREERS AVAILABLE

Prospective technicians can choose from a wide range of career options. Some of the major categories of employment involving repair and installation include the following:

- repairing or servicing electronic equipment used for entertainment in the home;
- installing, servicing, and repairing electronic equipment used in industry;

- repairing and servicing computers and related equipment;
- installing and repairing heating, refrigeration, air conditioning, and ventilation equipment;
- installing or servicing heavy equipment and various mechanical devices used in industry; and
- installing, servicing, and repairing various devices and systems ranging from alarm systems to vending machines.

These and other installation and repair fields share common characteristics. At the same time, each occupation has its own unique features. For details about of each these fields, see the chapters that follow.

JOB TITLES

The general area of installation and repair is a far-ranging one, with many different occupational titles. Some of the job titles used by the U.S. Department of Labor in this area are:

Air conditioning installer-servicer
Audio-video repairer
Cable splicer
Cash register servicer
Electrical appliance mechanic

Electromechanical technician
Electromedical-equipment repairer
Electronic-communications technician
Electronic-production-line maintenance mechanic
Electronic sales-and-service technician
Electronics mechanic
Electronics technician
Electronics utility worker
Electronic technician, nuclear reactor
Elevator constructor
Heating equipment installer
Industrial machinery mechanic
Instrument mechanic
Instrument repair supervisor
Line installer-repairer
Maintenance mechanic
Millwright
Millwright apprentice
Millwright supervisor
Musical-instrument repairer
Power saw mechanic
Radio mechanic
Service technician
Small-engine mechanic
Telephone installer-repairer
Vending machine servicer-repairer

JOB DEMAND

There is continuous demand for skilled workers who can install, service, or repair equipment. Employers need trained personnel in this area, and both businesses and individuals will pay excellent fees for the servicing or repair of items they own. After all, things break. Many types of equipment are too expensive to replace when they malfunction. Instead, the preferred option is to have equipment repaired. And, with today's increasingly advanced equipment, in most cases only specially trained technicians can do the job.

In addition, the growth in public concern about the environment is a positive factor. Many consumers would rather repair existing items than add unnecessarily to the "throw-away" mentality that has been so prevalent in the recent past.

From the viewpoint of career stability, installation and repair occupations offer a special advantage. While each job specialty requires certain knowledge, the basic techniques involved in installing or repairing equipment are quite similar. This means that skills learned in one job are readily transferable to other areas. Skilled, reliable workers in any career area in this field enjoy the potential of moving to other installation and repair fields if the need or desire arises.

QUESTIONS TO ASK YOURSELF ABOUT A CAREER IN INSTALLATION AND REPAIR

Does a career in this area sound appealing? If so, ask yourself the following basic questions:

- Do I enjoy working with my hands?
- Am I good with tools?
- Am I curious about the workings of engines, electronic equipment, or other devices?
- Do I like to fix things?
- Am I patient?
- Am I willing to go to school or complete on-the-job training to learn special skills?
- Am I good at math (or can I improve my math skills)?
- Do I have good eyesight (or eye problems that can be corrected with glasses or contact lenses)?
- Can I picture myself spending forty hours a week or more working with specialized equipment?
- Can I be a dependable, precise, dedicated worker?

If you can answer most or all of these questions positively, a career in installation and repair may be in your future. The remaining chapters provide an overview of a number of related occupations. As you read on, perhaps you will begin to focus on a career that seems just right for you.

CHAPTER 2

REPAIRING ELECTRONIC EQUIPMENT

With the growing importance of electronics in virtually all aspects of life, an increasingly promising career field is the repair of various types of electronic equipment. Some workers in this area specialize in repairing entertainment equipment and other items used primarily in the home. Others install or repair electronic devices used in industry. Some focus on specific types of electronic equipment such as computers, office machines, or telephones.

ELECTRONIC HOME ENTERTAINMENT EQUIPMENT REPAIR

People who enjoy working with electronic equipment can fulfill their career interests by repairing or servicing home-based electronic equipment. This field offers both solid career potential and a variety of work settings.

The typical family owns a number of electronic devices, including televisions, stereo systems, clock radios, videocassette players, compact disc players, microwave ovens, and so on. Since such devices are composed of electronic components that the average untrained person is unable to service or repair, a continuing demand exists for skilled technicians to provide such services.

Unlike many other mechanical devices, those mostly consisting of electronic components cannot be easily repaired by the average person. While someone with basic mechanical skills can take care of many aspects of servicing an automobile, for instance, the same is not true of a television set or videocassette recorder. Instead, the assistance of a specially trained technician is usually needed. Because of the complexity of such equipment and the large number of items that consumers purchase for home use, a substantial demand exists for technicians in this field.

Job Duties

Electronic home equipment repairers may also be known as service specialists, service technicians, electronic technicians, or other similar titles. Whatever the designation, these specialists service and repair various types of electronic equipment that can be found in the home. In some cases, such technicians focus on a single type of equipment such as televisions or radios. In other cases, they work with a variety of equipment types.

Electronic home equipment repairers perform duties such as the following:

- cleaning equipment and lubricating moving parts;
- consulting manuals or diagrams;
- determining if individual electronic components are in working order;
- replacing defective parts;
- using electronic testing equipment;
- operating soldering equipment;
- using basic hand tools;
- operating advanced electronic tools and equipment;
- working with electrical power sources;
- repairing equipment on-site;
- transporting equipment that can not be repaired on-site to a repair shop;
- calculating costs of parts and labor and other costs related to equipment maintenance or repair; and
- maintaining good communications with customers and fellow workers.

In completing these and other related tasks, technicians must be levelheaded, persistent, and methodical. They must be able to think analytically, first identifying problems and then deciding on appropriate ways to repair them.

The work performed by technicians of home electronic equipment is, in many ways, similar to the work of technicians of advanced electronic equipment. Those who have experience or training in related areas may find it feasible to undertake work in this specialty area.

Employers of Service Technicians

Technicians who repair home electronic equipment hold jobs throughout the United States and Canada. The majority of technicians work in metropolitan areas. Since most modern homes include electronic equipment, the need for service technicians is a continuing one, with the potential for employment existing virtually anywhere.

Many technicians in this field hold jobs with large department stores that operate their own service departments. Others are employed by stores that specialize in television and video equipment, musical equipment, home appliances, or other types of electronic equipment. Some work for businesses that provide repairs and service but do not sell equipment.

Some technicians in this field own and operate their own repair firms. The prospect of being self-employed is one attractive feature that distinguishes this occupation from many other career areas.

Those who run their own businesses follow a number of different approaches. Some work alone. Others form partnerships with one or more fellow technicians, or hire employees to work under their supervision. They may work out of their home, or may establish a full-fledged business that includes a well-equipped shop facility.

Obtaining Training

Special training is needed to prepare for a career in this field. In some cases, this is provided by the employer through a special apprenticeship or an on-the-job training program. In such instances, a high school diploma is usually the minimum educational requirement. Those who have taken vocational classes in electronics may have an advantage in gaining employment. In either case, employers add to the individual's knowledge base by providing classes, training videos, one-on-one instruction by experienced workers, or other training.

Many workers in this field gain skills by taking courses offered by trade schools or two-year colleges. They may complete entire degree or certificate programs, or may take a few select courses to get started in the field.

Working Conditions

An advantage of working as a service technician is that work is completed indoors, in settings that are usually comfortable. Such settings might be repair shops where customers bring equipment for repairs or service. Or they might consist of people's homes, where technicians go to perform on-site repairs. Both types of settings tend to be free of environmental discomforts such as loud noise or exposure to bad weather.

The typical service technician works in conditions that include good lighting, controlled temperatures, and adequate space at tables or work benches. Sometimes, the setting is informal, with workers being allowed to talk or listen to music while working. Another advantage is that comfortable, informal clothing is the norm, although some companies require special shirts, smocks, or work uniforms. Those who are self-employed may set their own standards in these areas.

Technicians who make service calls to private homes experience great variety in working environments. A single week's work might involve visits to modest homes, huge houses in affluent areas, apartments, and other residences in various sections of an urban or suburban area. This work also involves interacting with people of widely varying backgrounds, interests, and politeness. Technicians who can communicate well and adjust to the different personalities of customers tend to be the most successful in this kind of work.

In some cases, physical strength and mobility are needed to fulfill job duties. Service technicians may need to carry or move large items such as console television sets or other heavy equipment. Those with physical disabilities tend to work in shop settings rather than in the field.

Skills Needed

Technicians who specialize in repairing home entertainment equipment need most or all of the following skills:

- mechanical aptitude,
- knowledge of or ability to understand basic principles of electronics,
- good math skills,
- ability to analyze and solve problems,
- safety consciousness,
- skill in working with different types of tools,
- good oral communication skills,
- ability to maintain written records, and
- motivation to work consistently and patiently in providing service or repairs.

COMMERCIAL AND INDUSTRIAL ELECTRONIC EQUIPMENT REPAIR

Another occupational area in electronics is the installation and repair of equipment for businesses, manufacturing plants, the military, or other organizations. Technicians who specialize in this area are referred to as commercial and industrial electronics technicians. They may also be known as industrial-electronics technicians, electronics technicians, or some other variation of these job titles.

Technicians in this area work with a wide range of electronic equipment. For example, some work with industrial robots. Some focus on the installation and repair of electronic components of radar systems at commercial or military airports. Others work with industrial manufacturing equipment, communications equipment in the radio or television industry, high-tech medical equipment, or various other types of electronic devices.

Tasks Performed

The tasks undertaken by technicians in this field are similar in approach to those involved in working with home-based equipment. The major difference is that the equipment is used primarily in business, industry, or the military. As a result, some equipment is more complex in nature. In many cases, individual pieces of equipment function as part of an overall system, meaning that repair and installation tasks must be coordinated with other elements of such a system. In addition, equipment items may be larger or more comprehensive than those found in the home.

Typical job duties of industrial electronics technicians include the following:

- reading blueprints and manufacturers' specifications;
- using electronic devices such as oscilloscopes, signal generators, and voltmeters to test faulty equipment;
- diagnosing causes of equipment malfunctions;
- replacing defective components and wiring;
- adjusting mechanical parts with the use of hand tools;
- aligning, adjusting, and calibrating equipment;
- calibrating testing instruments;
- maintaining records of repairs, calibrations, and tests;
- installing equipment in industrial or military establishments; and
- operating electronic equipment.

Working Environment

Technicians install or repair electronic equipment in a variety of settings. They may perform their work at a manufacturing plant, military base, hospital, small business, or other facility. In some instances, the work may be performed on-site such as on an assembly line where an industrial robot is located, or inside a malfunctioning airplane. In other cases, technicians work in shops designed for equipment repair.

For the most part, technicians in this field enjoy comfortable working conditions. Part of this is due to the delicate nature of electronic devices. Since such equipment is often sensitive to extremes of humidity or temperature, it is often located in areas where the environment is carefully controlled. At the same time, the nature of repair work can be somewhat informal, with technicians wearing comfortable clothing and enjoying a relatively quiet, stress-free working atmosphere.

Of course, work done on-site will vary in terms of working conditions. In some cases, technicians may be exposed to loud noise, dirty manufacturing facilities, or other environmental challenges.

Skills and Aptitudes Needed

Men and women who plan to work as electronic technicians in the industrial or commercial sector should demonstrate most or all of the following traits:

- take pleasure in analyzing problems,
- enjoy working with their hands,
- possess good math skills,
- are patient and dependable,
- have good eyesight (or vision that can be improved with eyeglasses or contact lenses),
- are curious about technical matters, and
- are willing to learn new knowledge and skills.

Educational Preparation

Generally, some type of special training is needed to prepare for a career in industrial electronics repair. This can consist of options such as the following:

- vocational classes in electronics offered at the high school level;
- adult education programs taught by public-school districts;
- electronics classes offered by career colleges, trade schools, or proprietary schools;
- military training; and
- apprenticeship or on-the-job training programs.

One of the best ways to prepare for success in this area is to complete an electronics program in a two-year college. Many such programs are offered by community, junior, and technical colleges. For example, Johnson County Community College in Overland Park, Kansas, offers a two-year degree program in electronics technology. Students can specialize in particular areas of interest such as communications or medical electronics, but must complete a basic set of courses in electronics and general education. This consists of the following:

First Semester
Introduction to Electronics
Digital Electronics I
Computer Applications in Electronics

Composition I
Technical Mathematics I
Humanities and/or Art Elective

Second Semester
Circuit Analysis I
Digital Electronics II
Technical Writing I
Technical Mathematics II
Programming Elective

Third Semester
Electronic Devices
Circuit Analysis II
Microprocessors
Technical Physics I
Technical Electives
Health and/or Physical Education Elective

Fourth Semester
Electronic Devices II
Microcomputer Maintenance
Personal Communication
Social Science and/or Economics Electives

To earn an associate degree, students must complete 67 credit hours. The term "elective" refers to courses that students may select according to their own particular interests. Johnson County's technical electives in the electronics curriculum include the following:

PC DOS
PC DOS Intermediate
Local Area Network Fundamentals
Local Area Network Components
Local Area Network Operating Systems
Local Area Network Applications
Programmable Controllers
Introduction to Telecommunications
Advanced Programmable Controllers
Telecommunications
Medical Electronics Principles
Medical Electronics Applications
Electronics Internship I
Electronics Internship II

While this program is fairly representative of an electronics program in a two-year college, most schools have their own individual electronics curriculum. Many offer certificate programs, which differ from degree programs in that students take fewer courses and concentrate more on technical content than general courses. The typical certificate program can be completed in a year or less.

COMPUTER MAINTENANCE AND REPAIR

One of the most important of all technical occupations is the repair and maintenance of computers. After all, it is difficult to overstate the importance of computers in the contemporary business world, as well as in other areas of society. Think of virtually any organization—airports, hospitals, banks, government agencies, and so on—and then picture what happens if its computers fail to function properly. When computers go down, activity grinds to a halt. Without fully functioning computer systems, most organizations cannot keep up with the rapid pace and high volume of information involved in modern business transactions.

Even though computers are becoming easier to use, the average person would be hard-pressed to service or repair these complex electronic devices. Instead, such work is performed by computer repair technicians. Men and women filling this role may also be called computer service specialists, service technicians, or other similar job titles. They perform computer servicing and repair work for businesses, individuals, and other computer users.

Tasks Performed

Persons employed in computer maintenance and repair provide a variety of tasks. In general, computer repair technicians perform three basic functions:

1. They install computers and related equipment;
2. they provide routine maintenance of computer equipment; and

3. they diagnose problems, install replacement components, or otherwise repair computers and peripherals.

A representative list of competencies expected of technicians in this field can be found in the catalog of Northeast State Technical Community College in Blountville, Tennessee, which offers a training program for what it calls "microcomputer specialist technicians":

- analyze a problem and determine if it is solvable by a computer;
- describe the uses and characteristics of peripheral equipment available in a basic computer system;
- perform periodic preventive maintenance on computer peripheral equipment;
- recommend a hardware or software solution for a given problem;
- analyze, design, code, test, and document a program for use on a microcomputer;
- diagram a data communications network;
- install and configure commonly used device drivers; and
- ascertain general compatability requirements for hardware and software, and state specifications for use by management.

Working Environment

Technicians who work with computers enjoy good working conditions. As with others who deal with delicate electronic equipment, they usually work in indoor settings with good lighting and controlled temperatures.

Repair and servicing of computers may take place in a shop set up specifically for this purpose, or it may be conducted in offices and other settings where computers are located. Technicians often take computer components with them when making service calls, repairing equipment without removing it from the place where it is used, when possible. In some cases, this means traveling to businesses or other sites where computers are located.

Some technicians perform their work as service personnel of computer manufacturers. Many work directly for various types of firms that use computers. For example, a university may employ its own staff of technicians to maintain the many computers used for administrative, faculty, and student computing.

Obtaining Training

As with other installation and repair areas, working with computers requires special training. This is one area where it would be quite difficult to become self-taught, or to learn only through experience. Most people acquire skills in this field by participating in a structured training program.

IDENTIFYING THE RIGHT COMPUTER AREA

Before considering training options, it is important to distinguish from among the various career fields related to computers. After all, there is probably no "hotter" occupational area than the general vocation of working with computers. As a result, all kinds of training programs are available that have been designed in response to the growing importance of computers. Many of these programs, however, have nothing to do with the technical side of installing, repairing, or servicing computers. Instead, they focus on the use of computer equipment or software.

The field of computer operations, for example, deals more with the use of computers than their internal components. Similarly, computer programming is another distinct area where the work involves writing and working with computer programs as opposed to hands-on mechanical or electronics tasks. In considering computer training options, be sure to be aware of such distinctions.

COMMUNITY COLLEGE PROGRAMS

Many prospective computer technicians acquire their skills by studying at a community or technical college. Most such programs can be completed in one or two years of full-time study.

Ocean County College in Toms River, New Jersey, offers a good example of this type of instruction through its program in Computer Systems Testing and Servicing. This is a certificate program rather than an associate degree program, meaning it can be completed in about a year of full-time study compared to the two years needed for an associate degree.

Students who enroll in this program complete the following courses:

Basic Electricity
Instruments and Measurements
Basic Electronics
Computer Programming
Logic and the Microcomputer
Microcomputer System Maintenance and Troubleshooting
Computer Organization
Basic Composition
Intermediate Algebra I
Intermediate Algebra II

As this list of courses illustrates, the bulk of courses relates directly to computers or electronics. In addition, however, students master basic skills in algebra, since an understanding of mathematics is basic to this field; as well as an introductory English course, since everyone needs to develop good communication skills.

In terms of technical content, the following course descriptions typify the material covered.

Instruments and Measurements (two semester hours of
     credit)
     Students will learn to operate standard analog and
     digital laboratory instruments including the VOM, os-
     cilloscope, audio generator, and frequency counter.
     They will be able to solder PC board connections for
     IC chips, sockets, and standard components. Students
     will be able to collect data and display the data using
     proper graphing techniques on appropriate graph pa-
     per. The student will be able to use the scientific
     calculator, utilizing the majority of the scientific func-
     tions on the calculator.

Microcomputer System Maintenance and Troubleshooting
     (three semester hours of credit)
     With microcomputer hardware as a base, the student
     will become familiar with the architecture, functional
     operation, communications, I/O, maintenance, and re-
     pair of peripheral devices. A study of such peripheral
     devices will include printers, video terminals, and
     disc drives. Laboratory experience will emphasize
     hands-on exploration of the mechanics and circuitry
     that enable the devices to operate. Preventative rou-
     tine maintenance procedures will be stressed, along
     with troubleshooting techniques to distinguish be-
     tween malfunctioning hardware operation and soft-
     ware problems.

OTHER EDUCATIONAL OPTIONS

Trade schools, business colleges, and other non-collegiate schools also offer training in computer servicing and repair. For a list of representative schools, see Appendices B and C.

Chapter 5 covers educational options in more detail.

OFFICE EQUIPMENT REPAIR

Modern offices depend on a variety of advanced equipment. In addition to computers, just some of these items include

- dictation equipment,
- typewriters,
- calculators,
- copiers,
- paper shredders,
- printers,
- facsimile (fax) machines, and
- electronic postage meters.

When these and other items need service or repair, the work of special technicians is required. Office equipment repairers meet this need. They clean and service equipment that is in good working order, repair malfunctioning equipment, and sometimes set up new equipment.

Tasks Performed

Since most modern office equipment contains electronic components, technicians in this area tend to function in a way quite similar to those who repair home or commercial electronic equipment. The main difference is that they tend to specialize in one or more varieties of office machines.

Typical tasks performed include

- cleaning and adjusting equipment,
- diagnosing equipment malfunctions,
- replacing defective parts,
- adding enhancements or upgrades to existing equipment,
- developing links between individual equipment items in support of interoffice or intraoffice systems,
- dealing with equipment owners and users in a customer-service capacity, and
- completing reports and financial paperwork.

Many technicians in this field are employed by equipment manufacturers or distributors. They may also work for office supply stores or discount stores, or may be self-employed.

Typically, technicians in this field provide on-site service and repairs to office equipment. This means traveling to businesses and making service or repair calls. In most cases, such travel is restricted to a local geographical area, although with some highly specialized equipment more extensive travel may be required. Thus, the work involves customer relations, as well as technical services.

Training Options

Most workers in this field receive some type of on-the-job training. This is often provided by manufacturers for those who specialize in servicing or repairing their particular brands of equipment. Completion of electronics courses or other related courses in a vocational school or two-year college also can provide a basic background for working in this field. See Chapter 5 for more details.

TELEPHONE INSTALLATION AND REPAIR

The telephone is a staple of contemporary life. Most private homes have at least one telephone, and many have several. For businesses, life without telephones is virtually unthinkable. People use telephones to order goods, schedule meetings, convey important messages, and so on.

The high volume of telephone equipment and the importance it holds in a modern society lead to a continuing need for technicians skilled in working with telephone equipment. Telephone installers and repairers meet this need.

Such workers also may be called service technicians, station installers, field technicians, or telephone repairers.

Most telephone installers and repairers work as employees of telephone companies. They represent these companies in the field by installing, servicing, and repairing telephone equipment. This can include switchboards and multiunit telephone systems, as well as individual telephones.

Tasks Performed

Typical tasks technicians in this field perform include the following:

- traveling by van or truck to homes or business sites to provide installation or repair services;
- assembling telephone equipment;
- installing or relocating phone units;
- adding phone extensions and/or wiring;
- connecting telephones to outside phone lines;
- testing equipment to determine if it works properly;
- adjusting or replacing malfunctioning components;
- dealing with customers through oral and written communications; and
- processing records related to service calls.

Working conditions vary widely in this field. Technicians sometimes work outdoors, encountering different types of weather conditions. They may have to climb ladders or poles, crawl inside attics or storage spaces, or perform repairs in difficult-to-reach locations. On the other hand, much of their work is performed within the comfortable environments of homes or offices.

Background and Skills Needed

Most telephone companies require at least a high school diploma for employment in this area. Those who have studied electronics in vocational schools, or have completed related training at a two-year college, will have an advantage over other job applicants. While employers often provide some on-the-job training, they tend to prefer new employees who have already had related experience or training.

A special requirement in this field is the ability to distinguish colors, since telephone wiring is color coded. Good math skills and the ability to learn about the theory and practice of electronics are also helpful, given the increasingly "high tech" nature of telephone equipment.

VARIED CAREER OPTIONS

Thanks to the widespread reliance on electronic devices, technicians who work with electronic equipment enjoy a wide range of career options. Not only can they choose from the areas previously discussed in this chapter, but they can specialize in other specific areas of electronics.

For instance, the field of bio-electronics, or medical electronics, offers growing potential. Similarly, electronics technicians might concentrate in specialized areas such as radio and television or military weapons systems. With a continuing reliance on electronic components and devices, the demand for technicians trained to deal with such equipment provides a great deal of career potential.

CHAPTER 3

REPAIRING AND INSTALLING CLIMATE CONTROL EQUIPMENT AND INDUSTRIAL MACHINERY

Many installation and repair jobs focus more on working with mechanical equipment than with electronics, although some knowledge of electronics also may be needed in these areas. A typical example is climate control technology, which deals with air conditioning, heating, refrigeration, and ventilation equipment. The work of millwrights and industrial equipment repairers also shares similar characteristics.

CLIMATE CONTROL TECHNOLOGY

The ability to control temperature is one of the hallmarks of a modern technological society. Such technology allows us to keep homes, offices, and schools warm in the winter and cool in the summer. Beyond considerations of basic personal comfort, this kind of technology also helps meet fundamental needs for health and well-being. The ability to keep food and medicines cool, for example, prevents spoilage and prolongs the time they are still usable.

Refrigeration makes it possible for people on the East Coast of the United States or Canada to enjoy meat, fruits, and vegetables from California. At the same time, diners in Vancouver or San Francisco may partake of fish taken from the North Atlantic or the rivers of Brazil.

Regardless of geographical location, people may take advantage of blood needed for transfusions, antibiotics stored over a period of weeks, and other medical supplies that would not last without refrigeration.

The widespread existence of heating and cooling systems, and the importance people place on them, means there is a demand for technicians who can install, service, and repair equipment. These workers are known as heating, air conditioning, and refrigeration mechanics or by a number of other job titles. For example, some technicians such as furnace installers or heating equipment installers are known by the specific tasks in which they specialize.

Tasks Performed

Some of the tasks that technicians perform in this area include the following:

- using a variety of basic tools;
- reading and interpreting blueprints or other diagrams;
- examining malfunctioning equipment and identifying sources of problems;
- removing and replacing defective or worn-out parts;
- providing basic maintenance such as cleaning and lubricating equipment;
- developing plans for installing equipment used for refrigerating, cooling, or heating;
- installing individual equipment items or complete systems;
- working with different types of equipment, including heat pumps, boilers, electronic controls, pneumatic controls, compressors, and air flow ducts;
- following special safety procedures;
- writing business or technical reports;
- developing cost estimates; and
- communicating with customers.

Depending on individual job roles and training, some technicians also take on more complex responsibilities. For example, workers might operate microcomputers that are used in automated energy-management systems. They might develop blueprints or other complex drawings. They might also calculate heat load requirements or cooling capacities, or make other multifaceted calculations. Some supervise the work of others.

All technicians in this field do not perform the same duties. Some workers specialize in air conditioning equipment, but do not deal with furnaces. Others work exclusively with equipment that utilizes natural gas. Some workers, especially those with limited training or experience, perform a narrow range of tasks. For instance, they might install duct work to carry heated or cooled air from a central unit to various parts of a building, but might not work with compressors, electronic controls, or other complex equipment. Other technicians perform a wide range of complex tasks while working with different types of equipment.

Need for Workers

One advantage to a career in air conditioning and refrigeration is the relative stability of the field. Unlike some repair areas, economic ups and downs have relatively little impact on the need for services in this area. In times of a slow economy, some people tend to defer purchase or repair of some types of equipment (such as electronic entertainment items), but this is less true of climate control technology. If a large office's air conditioning system breaks down in the middle of July, for instance, those affected will waste no time in calling on qualified workers to address the problem. The same is true of heating systems that malfunction during cold weather, or of a grocery store refrigeration system any time of year.

Workers who specialize in the installation rather than the repair of this type of equipment are more subject to economic changes. When the construction industry slows down, it can have a negative effect on job prospects in this field.

Training Options

One way to obtain training in this field is to complete studies at a community or technical college. Many two-year colleges offer associate degree (two-year) programs or certificate (one-year) programs in climate control technology. Most still use the more traditional terminology of air conditioning, refrigeration, and heating systems, covering some combination of these three areas.

An example is Brevard Community College in Florida. Here, students may earn an associate degree in the technology of air conditioning, refrigeration, and heating systems by completing 68 credit hours of courses, including the following:

Introduction to Residential/Commercial Systems
Residential Load Calculations
Electricity for HVAC

Air Conditioning and Refrigeration Controls
Residential Heating Systems
Residential Heat Pump Technology
Mechanical Systems
Commercial Refrigeration I
Commercial Refrigeration II
Commercial Load Calculations
State and Local HVAC Codes
Modern Ventilation Systems
Commercial Heating Systems I
Commercial Heating Systems II
HVAC Trouble Shooting and Repair
Water Towers, Condensers, and Evaporators

Students who complete these courses, along with a few courses in basic areas such as communications, earn an associate degree. This prepares them for jobs as service technicians and system installers, among other positions.

A less demanding option consists of completing ten courses in this field for a total of thirty credit hours to receive a certificate rather than a degree. Graduates of this program pursue jobs as paraprofessionals working for contractors, dealers, and other commercial businesses.

Another option is to obtain on-the-job training. Some employers provide such training on an informal basis, especially for less complex work such as duct installing. A more formal approach is to complete an apprenticeship program. Such programs often are offered by labor unions and employers in a cooperative effort. Most take four years to complete, combining on-the-job learning with classroom instruction. Typical subjects workers learn through such programs include

- basic concepts of refrigeration and air conditioning,
- use of tools,
- safe working practices,
- blueprint reading,
- heat pump operations,
- refrigeration in the domestic setting,
- refrigeration in the commercial or industrial setting, and
- service and start-up procedures.

See Chapter 5 for more details on training options.

Working Conditions

Working conditions in climate control technology vary widely. Understandably, comfort levels tend to reflect the kind of work being done. If an air conditioning system has broken down during a summer heat wave, the working environment will undoubtedly be hotter than desired. A similar situation would apply during the winter if a furnace is not working or if climate control equipment is being installed in a new building.

At the same time, most work is done indoors. Thus, technicians tend to be sheltered from the full force of harsh weather. Indoor conditions will vary from damp, moist basements to fresh, newly constructed facilities. In some cases, workers may find it necessary to crawl under floors, in attics, or other hard-to-reach places as they install duct work or repair equipment.

On the plus side, the variety of working conditions may be seen as an advantage. Instead of working in the same environment every day, most climate control technicians travel from one work site to another. This may vary from industrial facilities or businesses to apartments and private homes. With each setting having its own unique characteristics, most technicians enjoy the experience of adjusting to the different working environments to which their job exposes them.

WORKING AS A MILLWRIGHT

A challenging career primarily dealing with equipment installation is that of the millwright. Millwrights install heavy equipment and machinery used in industry. They also dismantle such machinery when it is no longer needed.

Sometimes, they service or repair equipment once it has been installed.

Nature of Work

Millwrights provide the connection between the manufacturers of equipment and the workers who use it. They unpack equipment that has been transported to an industrial site and inspect it for damage or missing parts. They then move the equipment into place so that it can be used. With large pieces of machinery, this can be a slow and complicated process.

For example, a factory may receive a new machine that will be part of an assembly-line production process. Shipped on a railroad car, the machine weighs several tons. In this case, millwrights assist a crane operator in lifting the equipment off the truck and placing it inside a production facility. They then hook it up with existing equipment so the machine can become an integral component of the assembly line. Before considering the job complete, they conduct a trial run to make sure it operates properly.

Millwrights perform such functions for huge pieces of machinery, as well as smaller equipment items. In some settings, they also provide general repair and servicing functions for existing equipment. In other settings, the latter work is performed by industrial equipment repairers or other technicians.

Tasks Performed

Typical tasks performed by millwrights include the following:

- unloading and unpacking equipment;
- inspecting equipment for completeness and lack of damage;
- moving heavy equipment with the assistance of fork-truck operators or crane operators;
- preparing foundations for equipment sites with concrete, steel, wood, or other materials;
- using cables, ropes, crowbars, dolleys, and other means of moving equipment;
- using various hand and power tools, welding machines, lathes, and other equipment to prepare and assemble machinery;
- making measurements and mathematical calculations;
- reading blueprints and diagrams;
- dismantling machinery that is no longer needed; and
- servicing and repairing equipment.

These tasks are representative of typical work performed by millwrights. Actual work assignments vary with the type of employer and the nature of the equipment being installed.

Some millwrights specialize in just a few types of machinery. For example, some work for manufacturing companies that produce specialized machinery. As representatives of their companies, they travel to sites where customers have purchased equipment, installing the equipment and preparing it for use.

Other millwrights are employed by a single factory where they install any new equipment brought in from outside, dismantle old equipment, and service existing machinery. For instance, a company that makes automobile parts may employ millwrights to install and maintain its production equipment.

Some millwrights work with advanced equipment such as industrial robots. As technology advances, more of their duties deal with "high-tech" machinery.

Working Conditions

The working conditions of millwrights may differ substantially from those of some of the other occupations concentrating on repair and installation. The main differences is that instead of a quiet shop setting or the comfort of private residences where many repairs are made, millwrights work most commonly in manufacturing plants or other industrial settings. This means that they may face environmental challenges ranging from the noise of heavy machinery to exposure to dust, industrial chemicals, insulation, or other materials.

In many cases, millwrights perform their work outdoors. The process of unloading machinery from a truck or train car, for instance, usually starts outside of a building. Equipment such as pumps and generators also may be installed out of doors. As a result, millwrights tend to experience a wide range of weather conditions. Working outdoors may be a pleasant experience on a warm September day, providing an advantage that many other workers do not receive. On the other hand, struggling to install equipment during a rainy March day can be a muddy, wet, frustrating experience. Similarly, facing subfreezing temperatures during the middle of the winter can be an uncomfortable challenge.

Experienced millwrights learn to adapt to such situations and accept them as part of the job. They compensate for uncomfortable weather by dressing properly and, when possible, scheduling work around weather conditions. While such measures depend on individual preferences and levels of adaptability, many millwrights enjoy the variety they find in working conditions.

Getting Trained

In most cases, a high school diploma or the equivalent is the minimum educational level needed to work as a millwright. In addition, special training is usually needed to acquire the specific skills necessary in this occupation.

Most such training takes place on the job. The typical path is to start as a helper and then progress to full millwright duties and status. This may be done informally or through a formal apprenticeship program. The latter usually takes four years of classes and on-the-job experience. Typical subjects studied during this time include the following:

- basics of moving equipment,
- erecting and dismantling equipment,
- repair techniques,
- safety procedures,
- basic welding, and
- carpentry basics.

INDUSTRIAL MACHINERY REPAIR

Although millwrights sometimes repair existing equipment, this work is usually reserved for another type of specialist: the industrial machinery repairer. Men and women who perform this work also may be called industrial mechanics, industrial maintenance specialists, or maintenance technicians.

Tasks Performed

Most workers in this field are employed in manufacturing plants or other facilities where they provide basic service and repair functions. Typical job tasks include the following:

- inspecting machinery;
- providing basic service such as lubrication of moving parts;
- adjusting gears, bearings, and other mechanical components;
- identifying sources of equipment malfunctions and correcting them;
- reading and using repair manuals, blueprints, and other written materials;
- testing equipment to make sure it is working properly; and
- using electronic testing devices, hand tools, and other tools.

Specific duties vary depending on the place of employment and the types of equipment used in various industries. As with millwrights, some duties in this field are becoming increasingly complex as technology advances. At the same time, some methods for diagnosing problems and making repairs are becoming simple, thanks to advances in equipment used for such purposes.

Diverse Job Experiences

A special advantage of work in this field is that it offers variety in job tasks. In most cases, industrial equipment repairers, industrial mechanics, and other technicians in this area work with a variety of equipment. Unlike some other workers (such as technicians who specialize in television repair, for one example), those employed in industrial settings tend to find great variety in their jobs. One day a technician may make mechanical repairs on a large pump. The next day she might deal with a set of delicate electronic controls. Later in the same day, she might be called upon to replace a broken conveyor belt in an important part of an assembly line. At different times, as older machinery is replaced by newer, more modern equipment, she may need to study written manuals to master new service and repair techniques.

The end result of such variety is that for many workers, job-related boredom is avoided. Instead, technicians can look forward to different challenges in any given work day.

Working Conditions

When possible, industrial machinery repairers perform their work in repair shops or other indoor locations. This is not always possible, however, and some outdoor work is often required. In a large manufacturing plant where several industrial machinery workers are employed, job tasks may be divided. Some technicians may work exclusively in a shop, repairing equipment that is brought in by other workers. Others may have the responsibility of moving about the grounds and working with equipment that is too large to be moved, or is linked inextricably to other equipment. More often than not, technicians perform in both types of settings. Smaller, portable items are taken to a shop area for repair, while larger, heavier machinery is serviced or repaired wherever it is located.

Overall working conditions in this field reflect the nature of most mechanical repair work. This means that at times, technicians get dirty. They routinely work with oil, grease, and other lubricants. Rust, dust, and other contaminants are an everyday fact of life. This is not a significant problem for most technicians in the field, who realize that such conditions are just part of the job in an industrial setting. But those who plan to work in this field need to be aware of the nature of the typical working environment for industrial equipment repairers.

Getting Trained

Training options for industrial machinery repairers are similar to those of millwrights. Most workers obtain on-the-job training, either on an informal basis or through a structured apprenticeship program. Several labor unions sponsor apprenticeship programs.

Another option is to complete relevant studies at a technical or community college. For example, New River Community College in Dublin, Virginia, offers a certificate program in Industrial Maintenance. This program prepares students to work in commercial, industrial, or institutional plants. Students learn a variety of skills such as performing electrical installations, working with basic welding equipment, and handling lathe and milling-machine operations.

Typical courses completed in this program (normally completed on a part-time basis over two years, for a total of 31 or 32 credits) include the following:

Machine Blueprint Reading
Survey of Technical Mathematics I
Welding Elective
Small Gasoline Engines
Electrical Elective
Human Relations
Industrial Communication I
Machine Shop Operation
Principles of Refrigeration I

JOB ADVANTAGES

An advantage of repair and installation jobs in industry is their specialized nature. Generally, only workers with special training and experience can perform successfully in servicing and repairing the complex equipment used in industry. As a result, employers recognize the need for women and men who can perform this important work. Workers in this field generally earn good wages and, in some cases, are less vulnerable to job cutbacks than many other industrial workers.

CHAPTER 4

WORKING IN OTHER REPAIR AND INSTALLATION FIELDS

The previous chapters have highlighted some of the more common career areas related to installation and repair. In addition, a number of other occupational fields involve similar work. With the diversity of industrial and consumer products that need to be installed, serviced, or repaired, the breadth of this overall field is significant.

Following is an overview of some other challenging occupations involving installation and repair.

ALARM SYSTEM INSTALLATION AND REPAIR

The growth of crime is an unfortunate fact of contemporary life. Robbery, vandalism, and other criminal acts are all too common not only in large cities, but also in smaller towns and rural areas. To protect against illegal entry into homes or businesses, alarm systems have become an increasingly common application of contemporary technology. Along with guarding against theft, alarm systems provide warnings about fires, smoke, or other environmental threats.

The nature of alarm systems ranges widely. Some are simple devices triggered by the opening of a door or window or by a sudden increase in temperature. Others are complex systems involving computers, laser beams, or other advanced features.

Work Performed

The special characteristics of alarm systems require the attention of technicians who can install, service, or repair this type of equipment. Technicians in this area may perform most or all of the following functions:

- inspecting existing systems to insure that they are working properly;
- training homeowners or business employees in proper use of alarm equipment;
- diagnosing problems in malfunctioning alarm systems;
- replacing alarm system components or making other repairs on-site;
- making equipment adjustments or repairs in a shop setting;

- using various tools ranging from wrenches or screw-drivers to computers or electronic testing equipment; and
- dealing with customers to determine problems and communicate solutions.

Alarm systems are used widely both in private residences and businesses. As a result, technicians in this area work in a variety of settings. For most workers, the nature of the job involves moving frequently from one site to the next, putting in only a few hours or days at any one location. This provides a great deal of job mobility and a certain amount of variety in daily routine.

Skills and Background Needed

The same general aptitudes are needed for alarm system work as in a number of other areas involving equipment installation and repair. Basic mechanical aptitude, the ability to work with tools, good math skills, and the capacity to learn the basics of electronics are all helpful. In addition, honesty and reliability are important. Clients place great trust in the individuals who work with alarm systems. Technicians who install or repair such equipment must do their jobs well, and must maintain an awareness of the need to protect the confidentiality of information about access to buildings and the valuable items they house.

Although some schools provide special training in this area, most alarm system repairers acquire their skills through on-the-job training. Many complete some type of training in electronics or a related field before seeking jobs in this area. Employers then provide training related specifically to alarm equipment. Typical employers include equipment manufacturers, firms that specialize in providing alarm systems for businesses, and companies that install and service home-based security systems.

ELEVATOR INSTALLATION AND REPAIR

A relatively small but highly important career area is that of the elevator installer and repairer. People who work in this field assemble, install, and replace elevators, escalators, and related equipment. They also may be called elevator mechanics or elevator constructors.

Work Performed

Some technicians in this area specialize in installing new equipment. Others concentrate on servicing and repairing existing equipment. Some perform both types of functions.

Typically, elevator installers and repairers perform tasks such as the following:

- preparing elevator shafts by bolting or welding steel rails to shaft walls;
- installing electrical conduits;
- installing electric wiring and electronic controls;
- assembling elevator cars;
- installing various components of elevator systems such as doors, winches, counterweights, and hydraulic cylinders;
- installing escalators and their components;
- installing powered walkways, material lifts, dumbwaiters, and related devices;
- troubleshooting equipment problems;
- making adjustments to recently installed equipment; and
- servicing existing equipment to prevent breakdowns.

A special feature of this occupation is a focus on safety standards. Since people's lives rely on the safety of elevators, workers in this area must pay a great deal of attention to thorough, safe practices of installation and repair. In fact, in some geographical areas, elevator installers and repairers must pass a licensing examination before they are allowed to work without direct supervision.

Skills and Background Needed

Most technicians in this field learn their skills through on-the-job training provided jointly by employers and the

International Union of Elevator Constructors. They start out as trainees or helpers, advancing to fully qualified technicians in about four years. Their training includes special classes, as well as supervised experience in actual job settings.

As with other fields involving repair and installation, workers in this area need a basic aptitude for mechanical concepts, the ability to work with tools, and the capacity to learn special techniques in working with specific types of equipment. They also need to be dependable and dedicated to doing a good job.

HOME APPLIANCE AND POWER TOOL REPAIR

Virtually every family in the United States and Canada owns or uses several types of home appliances. Machines such as refrigerators, washing machines, clothes dryers, automatic dishwashers, stoves, and microwave ovens make everyday living more efficient. In addition, power tools such as lawn mowers and electric saws and drills are common staples of home maintenance, as well as hobby efforts.

These and other mechanical devices found in the home share one thing in common: Their effectiveness is not permanent. Appliances and tools break down; yet their value may make it less expensive to repair them than to replace them. Thus, a continuing demand exists for technicians who can repair such items.

Repairers of home appliances and power tools meet this demand. They make repairs on malfunctioning equipment and perform preventive maintenance functions. Thanks to their work, the useful working life of basic equipment can be extended significantly.

Tasks Performed

Tasks performed by workers who repair home appliances and power tools vary according to the type of equipment being installed, serviced, or repaired. Work done on a gas stove, for instance, will differ in some respects from that for electric stoves. Some technicians specialize in repairing small appliances such as toasters, blenders, or vacuum cleaners. Some concentrate on large appliances such as washing machines, dryers, refrigerators, stoves, or freezers. Others work primarily with power tools such as gasoline or electric saws, electric dills, lawnmowers, and hedge trimmers. Many work with a combination of equipment items rather than specializing in a single area.

Examples of individual tasks these technicians perform include the following:

- travel by truck or van to home sites;
- identify causes of equipment malfunctions;
- remove equipment and transport it to a shop for repair;
- repair equipment on-site if it is too large to transport conveniently;
- consult manuals or wiring diagrams;
- remove worn or defective components;
- replace internal parts such as bearings, gears, or belts;
- replace circuit boards or other electronic parts;
- clean or replace components such as thermostats, valves, and heating elements;
- replace electrical wiring;
- add lubricants;
- make estimates of repair costs;
- keep records related to service calls, repairs, and costs of parts; and
- use various tools and testing devices such as voltmeters, screwdrivers, soldering guns, and wrenches, as well as tools designed for specific types of appliances.

Background and Skills Needed

The skills needed for this occupation are similar to those required in most other fields involving repair and installation. These include the following:

- an overall aptitude for mechanical work,
- skill in using different types of tools,
- the ability to identify malfunctions or problems and determine appropriate ways to correct them,
- good eyesight (may be corrected by optical wear),
- attention to detail,
- physical mobility, and
- the ability to derive satisfaction from working with one's hands and performing tasks of a mechanical nature.

In addition, other skills may be helpful. The ability to interact well with customers, in particular, is an important asset for those who travel to people's homes and make repairs there. This includes good oral communication skills, as well as skill in good customer relations (that is, interacting with customers in an effective and positive way). Technicians must be able to listen to customers report the nature of problems and then, in many cases, explain how they plan to eliminate them. Because some customers are talkative and others are reserved, technicians must be prepared to deal with a range of personality traits. An ability to communicate and get along with other people makes this occupation more enjoyable and enhances job performance.

The ability to work independently is another helpful asset. Home appliance and power tool repairers need to be task-oriented people who can work in a productive fashion without the need for a great deal of supervision. Other helpful attributes include

- good driving skills,
- the ability to follow directions in locating residences,
- the ability to calculate repair bills and handle checks or cash,
- flexibility in dealing with different brands or styles of equipment,
- patience in performing work to its completion,
- the ability to work quickly and efficiently, and
- a high level of commitment to doing a good job.

Getting Trained

Training options for this field are similar to those for technicians who repair home-based electronic entertainment equipment. This may consist of on-the-job training, completion of studies at a trade school or two-year college, or a combination of the two.

Many employers of technicians in this field provide training for employees. Department stores and appliance stores, for example, are among the most common employers of home appliance or power tool repairers. These businesses often provide training for new staff that may range from basic repair techniques to safety procedures. New employees are typically assigned simple tasks such as assisting experienced workers, cleaning equipment, or learning to dismantle and service a single type of tool or appliance. They then learn more complex tasks under the supervision of veteran technicians.

Many workers in this field obtain training from a school or college. Some vocational high schools, trade schools, and two-year colleges offer programs in appliance repair, small-engine repair or related fields. Many others offer courses in electronics or related fields that can be combined with on-the-job training to gain the necessary skills.

When hiring new staff in this field, most employers prefer applicants who have had some type of formal vocational training. If a full-scale program in appliance or power tool repair is not available, persons interested in this occupation might do well to complete, as a minimum, a few basic courses in electricity, gasoline-engine repair, or some related area.

In addition to their initial training, workers in this field often complete more training on a continuing basis. This may consist of noncredit classes, seminars, or more extensive courses. Appliance repairers who deal with appliances based on the use of natural gas, for instance, may undergo special training that emphasizes safety standards. Some states even require certification of workers who install or repair gas appliances.

VENDING MACHINE REPAIR

Ready for a candy bar or soft drink? Or how about a snack targeted more to the health conscious? If so, one quick source is the nearest vending machine.

The vending machine has become a popular fixture of contemporary life. In schools, hospitals, factories, and other facilities, such machines are common. Vending machines also can be found on street corners, at the exits of department stores, at highway rest stops, at gas stations, and so on. In fact, in all but the most rural areas, chances are you would not have to travel far to reach the nearest vending machine. For food, drinks, stamps, and various other commodities, vending machines provide convenience in obtaining consumer goods.

The popularity and accessibility of these machines means that they undergo frequent use, requiring regular service and repair.

Vending machine servicers and repairers keep machines in working order. They install new machines, service existing machines, and make repairs when needed. Some also complete related tasks such as stocking machines with items for sale.

Tasks Performed

Job duties in this field vary. Some employees work with a variety of machines, while others concentrate on a single type of machine such as beverage machines, coin changers, or microwave ovens.

Basic tasks performed by vending machine servicers and repairers may include the following duties:

- checking out new machines before they are placed in operation, including such steps as

  making sure beverages are mixed properly,

  seeing that refrigeration units work or that microwave units do not leak radiation,

  making certain that electrical or water connections are made properly, and

  seeing that springs, handles, plungers, and chutes work properly;

- making repairs when machines malfunction, including determining causes of problems,

  tightening loose wires or parts,

  replacing circuit boards or other defective parts, and

  transporting machines that can not be fixed on-site to repair-shop settings;

- performing preventive maintenance tasks such as cleaning, tightening, and lubricating parts;

- completing written service reports and orders for replacement parts;
- adding merchandise to machines to replace purchased items;
- collecting money from machines; and
- maintaining financial, merchandise, and service records.

Some vending machine servicers and repairers work with advanced electronic components, which are becoming increasingly common. Others work only with traditional mechanical components.

Many workers in this field spend a significant portion of their time traveling from one vending machine site to another. Some work exclusively in repair shops, relying on other workers to deliver and pick up machines and components. Many combine both types of work experiences, working in the field and in a shop setting.

Skills Needed

As with other repair and installation careers, workers in this field need skills such as the following:

- basic mechanical aptitude;
- good eyesight (may be corrected with glasses or contact lenses);

- manual dexterity;
- the ability to use different types of tools, including screwdrivers,
  hammers,
  wrenches,
  grinding wheels, saws, and other power tools, and electronic testing equipment;
- the ability to analyze and solve problems; and
- the ability to derive a basic sense of satisfaction from working with one's hands.

In addition, most jobs in this field require physical mobility. This may include the ability to drive vans or trucks, as well as to move machines and assume different positions while working on them. For persons with certain types of disabilities, jobs where work is performed in a shop rather than in the field may represent more realistic possibilities.

Getting Trained

Some type of special training is usually necessary to work effectively in this field. This may consist of informal training on the job, a formal job training program, or completion of courses offered by a trade school, technical school, or two-year college.

Many workers start out with no prior training related to vending machine service or repair. They serve as helpers of experienced workers, performing relatively simple functions such as cleaning or painting old machines. In time, they learn how to perform other more complex tasks.

Some workers gain an edge by completing courses in areas related to this occupational field. This may consist of a program designed specifically to prepare vending machine servicers and repairers. Or it may involve more general courses in electronics or a related field. For more details about educational options, see Chapter 5. Also, Appendices B and C provide names and addresses of selected schools and colleges offering programs in this area or related fields.

An alternative means of gaining skills in this field is to complete an apprenticeship program. For example, the National Automatic Merchandising Association sponsors a program that combines job experience with 144 hours of instruction each year for three years.

LINE INSTALLATION AND REPAIR

Go outside and look around you. Chances are you will see multiple lines running from poles into homes or businesses. Such lines, carrying electricity, television programming, or telephone messages, provide essential links for the movement of electrical power or communications. In addition, underground cables provide similar links on a widespread basis.

The installation, repair, and maintenance of lines and cables is an important occupational area. Women and men who provide such services hold several types of positions:

- Line installers install wires or cables and the poles or terminals on which they are placed. These specialists also may be called construction line workers or outside plant technicians.
- Cable splicers make connections in lines or cables. They connect lines or the individual wires or lines within them. Workers who perform these tasks also may be known as cable splicing technicians.
- Cable television installers and repairers complete the connections between cable television lines and television sets. They also are known as service technicians.

Skills and Background Needed

To work successfully in this field, people need similar skills and training as in other installation and repair careers. They need mechanical aptitude and the ability to learn about electricity and other technical information.

In addition, most line installers and cable installers need certain physical capabilities, including climbing without fear of heights, moving about with ease, and distinguishing colors, since wiring is often color coded.

Most workers in this field acquire their skills through on-the-job training. In addition to training in basic aspects of the job, they undergo special training in safety techniques. As technology advances, subjects such as working with fiber optics are playing an increasing role in job training, as well as practice.

OTHER INSTALLATION AND REPAIR AREAS

The previous chapters have provided an overview of some of the more visible occupational areas of installation and repair. Other related areas that may involve similar work include

- repairing boats, motorcycles, or small engines;
- repairing and tuning musical instruments;
- installing and repairing communications equipment;
- repairing automobiles, farm equipment, or aircraft;
- automotive body repair;
- repairing and installing military weapons systems; and
- installing and repairing optical, laboratory, or scientific instruments.

While each career area has its own distinct features, these and related areas share many characteristics. People who are suited to any one field may find they also have potential for other repair and installation career areas.

CHAPTER 5

OBTAINING OCCUPATIONAL TRAINING

A career in any occupation involving repair and installation requires specialized knowledge. For most jobs, this means completing a formal training or educational program. The typical employer either expects job applicants to have completed job-related training, provides special training for new employees, or both.

Many workers prepare for a job in this field by studying at a technical school, community college, or other school that offers related technical courses. Others are hired based on potential rather than acquired skills, learning the basics of the job through instruction provided by the employer.

EMPLOYER-SPONSORED TRAINING

One way to learn job skills is to participate in training activities sponsored by employers. In such cases, companies hire workers and then provide instruction in performing basic job tasks. For example, they may provide classes or seminars related to specific types of equipment and the appropriate techniques for working with them.

In some cases, new employees have no special background in the area in which they will be working. Employers hire them based on potential rather than job knowledge. In other cases, workers with related job training or experience are hired, and then given additional on-the-job training that focuses on the expectations of that particular company.

The approaches used in company-sponsored training vary. Some opportunities consist of very informal on-the-job training, where new workers start out with simple tasks and then learn by helping or observing more experienced workers. Other activities include seminars or structured classes. Such classes cover material not unlike that taught in schools, but that may be condensed over a shorter time period. Other training methods include instructional videos, simulation exercises, workbooks, computerized learning programs, and individualized study of training manuals.

A major advantage of company training programs is the fact that in many cases, no cost to the employee is involved. In fact, employees even may be paid to attend. Furthermore, initial training is highly focused and can be finished in a relatively short time.

Details about on-the-job training opportunities can be obtained from the personnel office of companies that hire technicians who perform installation and repair work. Prospective employees also may watch for announcements of job and training opportunities in newspapers.

APPRENTICESHIP PROGRAMS

Employers and unions sometimes sponsor special apprenticeship programs that provide structured job training. This approach allows workers with little experience to combine job experience with special training, all under the guidance of experienced workers in the area of training.

Apprenticeships have a long and distinguished history. They reached perhaps their greatest level of importance in Europe during the Middle Ages. Workers learned the special skills of a trade or craft by serving as apprentices to highly experienced workers. Such programs are less common today, but are still used by employers and trade unions in a number of fields related to installation and repair. They can provide a highly effective method of training.

Contemporary apprenticeship programs are offered by companies, labor unions, schools, or a combination of these entities. Most take several years to complete. During this time, participants earn wages and benefits. They are usually lower than those of workers who have completed apprentice training, but the potential to earn while learning still represents an advantage over regular school-based training. As workers gain skills and experience, their pay usually increases.

The Virginia Department of Labor and Industry lists the following benefits of apprenticeship programs:

- better qualifications,
- increased job satisfaction,
- potential for advancement to supervisory positions,
- preparation for licensing exams,
- better understanding of job-related theory, and
- evidence of skill attainment.

Apprenticeship is an efficient, proven training method. Such training offers the opportunity to learn from experienced workers while earning wages and maintaining employment. To learn more about apprenticeship opportunities, contact employers in your area, your state's apprenticeship training council, or a Federal office dealing with apprenticeship training. A list of Federal offices dealing with such training is provided in Appendix D.

SCHOOL AND COLLEGE PROGRAMS

Another way to prepare for an installation and repair career is to attend a school or college that offers related programs. This is an increasingly viable option given today's competitive job market and the complexity of modern equipment. Even for occupations where such training is not an absolute requirement, some degree of formal instruction often is worthwhile.

In pursuing such options, the following factors should be considered:

- types of schools offering appropriate programs,
- level of instruction,
- quality of offerings,
- length of time required for completion,
- how programs relate to expectations and requirements of employers,
- school location,
- costs, and
- availability of financial aid.

Choosing the Right School

Choosing the right school or college is an important step. In making such a choice, be sure to realize that all schools and programs are not the same. Schools vary widely in quality, types of programs, cost, and applicability to your chosen career.

Types of schools that offer training related to repair and installation careers may include the following:

- secondary-level vocational schools,
- trade or technical schools (sometimes called business colleges or career colleges), and
- community, junior, or technical colleges.

VOCATIONAL SCHOOLS

Those who have not yet completed high school may benefit from classes at vocational high schools or traditional high schools that offer vocational courses. These can range from full-scale programs in special areas such as small-engine repair or appliance repair to basic courses in electricity or electronics. Other courses such as blueprint reading, carpentry, or technical writing also might be offered. Usually, any instruction at this level will be potentially valuable for future applications.

Adults who are interested in such classes might consider adult education programs administered by public school districts or vocational schools. These programs often are offered at night to accommodate adult learners. An advantage is that adult education classes may be available at little or no cost.

TRADE AND BUSINESS SCHOOLS

Other training options are offered by postsecondary trade or business schools. These institutions may be called vocational institutes, proprietary schools, or technical schools. Some prefer to be called career colleges or business colleges, but they are not bona fide colleges. Instead of offering higher education, their main emphasis is on occupational training.

The main advantage of trade and business schools is that they focus on career studies and usually do not require extra courses in English, history, or other traditional subjects. As a result, programs can be completed relatively quickly. Students who do not particularly enjoy school may see this as an asset.

At the same time, most students find there are some disadvantages associated with business or trade schools. One is that they may be expensive. These schools operate as businesses rather than nonprofit enterprises. As a result, costs may be much higher than two-year colleges or public schools.

In some cases, higher fees can be offset by grants or loans. But students must be aware that they may end up with large sums of money to repay. Cost and financial aid should be considered carefully before a commitment is made.

Another possible disadvantage is that courses completed at trade or business schools are not transferable to colleges and universities. If you enroll at such a school and then later decide to attend a community college or four-year college, you will need to start over.

In considering schools of this type, watch for recognition such as accreditation by the Career College Association. A list of selected schools accredited by this group appears in Appendix B.

COMMUNITY, JUNIOR, AND TECHNICAL COLLEGES

Many two-year colleges, which may be called junior, community, or technical colleges, offer programs or courses related to installation and repair career fields. Some teach electronics, computer technology, climate control technology, and other programs related to installation and repair. Many teach individual courses, if not full-scale programs in such areas.

Although these schools are called "two-year" colleges, most also offer programs that can be completed in a year or less. These certificate-level programs are similar to those offered by trade schools in that courses in math or the liberal arts may not be required. Associate-degree programs, on the other hand, generally require two academic years to complete, since students must take some general studies courses, as well as technical classes.

An advantage of two-year colleges is that many of their courses can be transferred to four-year colleges and universities. This may be important if plans may change in the future to include a four-year degree.

In addition, community colleges tend to be much less expensive than either business schools or four-year colleges and universities. A year or two of full-time study can be affordable, high quality, and highly valuable in preparing for a job in a field related to repair and installation. A partial list of two-year colleges offering programs related to repair and installation is provided in Appendix C.

CHAPTER 6

SALARIES, WAGES, AND BENEFITS

Technicians who specialize in installation and repair generally earn good incomes. Actual salaries or wages vary from one field to the next and also are affected by type of employer, location, experience, and other factors. For the most part, however, women and men employed in this field earn substantially more than workers in many other areas, especially those lacking special skills or training.

According to the United States Department of Labor, mechanics and repairers as a group had a median income of $490 a week in 1991. This means that one-half of all workers made more than this amount, and one-half made less. On a yearly basis, the median income was $25,480. Many workers, especially those with a good deal of job experience, earned significantly more.

Median income for selected specific fields in the same time period was as follows:

- Telephone installers and repairers averaged $645 weekly, or $33,540 yearly.
- Telephone line installers and repairers earned an average of $629 weekly, or $32,708 per year.
- Electrical and electronic equipment repairers averaged $591 a week, or $30,732 annually.
- Industrial machinery repairers earned a median income of $491 weekly, or $25,532 yearly.
- Computer equipment repairers earned an average of $569 a week, or $29,588.

In these and other related fields, it is not uncommon for experienced technicians to earn $30,000 to $40,000 a year. Some, especially those in supervisory positions, earn more.

In general, people employed in repair and installation occupations earn good salaries and benefits. Their incomes are higher than those of many other occupations, particularly those where no specialized training or job knowledge is needed.

FACTORS AFFECTING SALARIES AND WAGES

A number of factors can affect the level of payment earned in various repair and installation fields. In addition to differences from one field to another, wages and salaries can vary within similar job categories. Factors influencing pay include the following:

Geography. Technicians in metropolitan areas tend to earn more than those in rural areas. The same is true in areas with a higher cost of living. For instance, housing, food, and other staples cost more in Chicago or San Francisco than in rural Alabama. As a result, workers in all categories, including repair and installation fields, usually earn correspondingly higher salaries and wages.

Level of Education. In many cases, jobs requiring training after high school pay higher salaries than would otherwise be the case.

The Economy. In times of inflation, wages and salaries usually rise. During recessions or times of slow economic growth, wages and salaries may be frozen or grow only a little.

Previous Experience. Highly experienced technicians usually earn higher salaries or wages than those with limited job experience. In many companies, a structured wage scale takes experience into account in setting pay levels.

Overtime. Technicians who work extra hours may earn overtime pay. This is usually one and one-half to two times normal hourly wages, depending on the circumstances.

Employer Status. New businesses may not be able to pay salaries that are as high as those of established firms. In the same vein, companies that hire only union members may pay higher wages than nonunion employers.

Competition. Competition among companies that employ technicians in similar types of positions may affect pay levels. If they need to compete with other firms to hire or retain technicians, employers will often increase basic pay levels.

BENEFITS

Wages or salaries are only part of a standard pay package. In addition, most employers provide several types of benefits to their employees.

Benefits vary from one employer to another. They are usually more extensive for full-time employees than for part-time workers. In many cases, the total amount paid in benefits totals twenty to thirty percent of base wages or salaries.

Typical benefits include the following:

- vacations,
- health insurance,
- retirement funds,
- sick leave,
- workmen's compensation in case of injury, and
- Social Security payments.

Also, some employers allow employees to participate in profit sharing programs. Employees may own stock in the company, or receive bonuses based on the overall performance of the firm. Other benefits may include bonuses for holidays or for performing exceptionally well or extra medical coverage such as dental or optical insurance.

In looking at any employment situation, benefits should be considered. Health insurance, for example, can be worth thousands of dollars. Other benefits also are important. They should be carefully evaluated during consideration of different employment options.

CHAPTER 7

PROFESSIONAL ORGANIZATIONS AND CERTIFICATIONS

A common facet of working life in the United States and Canada is membership in professional organizations. Many technicians who work in installation and repair fields belong to such groups. These may consist of professional societies, labor unions, or other organizations.

REASONS FOR BELONGING

Why do technicians belong to professional organizations? Most find that joining one or more professional groups is highly worthwhile. Members receive benefits ranging from the sharing of job-related information to the sponsorship of special programs for certifying special competencies.

The Professional Electronics Technicians Association lists these reasons technicians find membership helpful:

- improvement of pay and prestige through group efforts;
- monthly information on various phases of electronics, including circuit descriptions and short quizzes;
- business seminars and other management information for those who are service managers;
- discounts on trade magazines, technical books, and other items;
- opportunities for people with common interests to get together and share information; and
- a certification program.

Based in Greencastle, Indiana, this organization welcomes memberships from individual technicians, electronics students, and owners of businesses and institutions, among others. Annual membership fees range from $20 for students to $125 for institutions. Along with regular Electronics Technicians Association (ETA) membership, each member may belong to one of the following divisions:

- Educators
- Certified Technicians
- Canadian Division
- Communication Techs
- Medical
- Industrial
- Shopowners

A similar organization in another technical field is the Refrigeration Service Engineers Society, based in Des Plaines, Illinois. Members of this society are technicians in air conditioning and refrigeration or those with interests in the field. Over 29,000 members belong to nearly 400 local chapters. Some of the group's services to members include

- various training courses in air conditioning and refrigeration,
- publications related to the field,
- educational meetings,
- low-cost group insurance, and
- business-building aids such as business forms and engravings for stationery.

Another professional group is the International Society of Certified Electronics Technicians. It has more than 28,000 members. With national headquarters in Ft. Worth, Texas, this society provides the following:

- administration of a Certified Electronics Technician (CET) Program,
- a national apprenticeship and training program,
- technical information training and upgrading programs, and
- serviceability inspection programs.

Only those electronics technicians who have passed the organization's CET exam may belong to this organization.

Once they have become full members, technicians benefit from a wide range of information, networking with other members, and the professional status of having passed a recognized certification process.

Men and women in other fields related to repair and installation may participate in other professional groups. Those connected with the telephone industry, for instance, may belong to the United States Telephone Association. Persons employed in jobs other than those as technicians may elect to participate in other organizations. For example, owners of their own electronics repair firms might join the National Electronics Sales and Service Dealers Association. Other organizations of this type are listed in Appendix A.

UNION MEMBERSHIP

Many technicians belong to labor unions. Unions join workers together to promote the welfare of members of the group and strengthen their position in dealing with employers.

Labor unions have been a major force in American and Canadian business and industry since the nineteenth century. They have helped bring about many improvements in worker rights and benefits. Shorter work weeks, higher

pay, fringe benefits, and expanded job rights have all come about largely as a result of the work of labor unions.

Technicians in installation and repair fields may belong to a number of unions, including the following:

- Communications Workers of America
- International Brotherhood of Electrical Workers
- International Union of Electronic, Electrical, Salaried, Machine, and Furniture Workers
- International Union of Elevator Constructors
- United Electrical, Radio, and Machine Workers of America
- Federation of Westinghouse Independent Salaried Unions

Benefits of membership in a labor union can include

- training opportunities,
- guaranteed wage levels and other benefits,
- the right to vote for union officials,
- the right to vote about proposed union-management contracts,
- publications,
- retirement plans,
- health benefits, and
- protection against unfair labor practices.

To receive these services, members must pay dues. A representative amount for most union memberships is the equivalent of two hours' wages each month. Funds from members support union activities for improving pay and working conditions, as well as other activities such as lobbying for legislation that will benefit members.

Not all technicians belong to unions. Membership in unions is more common among employees of large firms than it is among employees of smaller companies. It also is more common in urban areas than rural ones and in certain geographical locations. For instance, unions are not as common in the southern United States as in some parts of Canada and the northern United States.

With some jobs, union membership is virtually required. In other cases, it is not feasible. In still other instances, the choice is up to the individual technician. In any event, many technicians in repair and installation fields find union membership worthwhile.

CERTIFICATIONS

In some fields related to installation and repair, technicians have the opportunity to become certified through a structured certification process.

For example, the Refrigeration Service Engineers Society issues diplomas for technicians who have completed specified training courses. Similarly, several electronics organizations provide certification opportunities.

The Electronics Technicians Association sponsors an examination process that leads to designations at the associate, journeyman, senior, or master level. Members also can become certified in special areas such as Consumer Electronics, Video Distribution, and Bio-medical Technology.

A similar certification process is sponsored by the International Society of Certified Electronics Technicians. This group sponsors examinations leading to two certification levels: associate and journeyman.

When possible, technicians are well advised to acquire certifications. Such credentials show employers, peers, and customers that technicians have a high level of proficiency that has been documented through the certification process.

CHAPTER 8

GETTING STARTED IN REPAIR AND INSTALLATION

After having reviewed the information in previous chapters, perhaps a career in installation and repair seems appealing. Perhaps, in fact, you can envision yourself working as a technician in one of the fields discussed in this book. If so, the next step is to develop plans for pursuing such a career and then implement them.

Following are some possible steps to take to get started with a career in an installation and repair field:

1. If you are a high school student, enroll in classes related to your field of interest. For instance, take math, physics, or vocational classes in electronics, small-engine repair, or other similar areas.
2. Along with high school courses, determine what kinds of training programs are available for training after high school.

3. If you decide to go on to school after graduation, complete admission applications and other required forms.

   Fill out forms to apply for financial aid, if needed.

4. Once in school, attend class regularly, apply yourself, and earn a diploma or degree in an appropriate technical field.

5. For an alternative approach, apply for a job with a company that has its own on-the-job training program or that cooperates in offering an apprenticeship program.

OBTAINING A JOB

After developing educational credentials or when you are otherwise ready to seek employment, the next step is to apply for a job for which you feel qualified. This may require some concerted effort on your part, but the end result can be the start of a career you can enjoy.

Finding Out About Job Vacancies

A good place to locate job openings is the classified section of any newspaper. Daily papers serving larger towns and cities are usually the best bet. The Sunday issue usually has the most extensive job listings.

Following is a representative classified ad for an installation and repair technician:

INSTALLATION & REPAIR TECHNICIAN
Northside Community College is accepting applications for an Installation and Repair Technician in the college's Computing Center. Duties: Assisting with installation and service of multiple-vendor microcomputers and associated peripheral equipment and maintenance and service of office and classroom electronic equipment. Applicants must have considerable training or experience in electronic servicing of microcomputers and peripheral equipment with the ability to diagnose and service problems down to the board component level and excellent oral communication skills. Training and experience in servicing of electronic and electromechanical office equipment, electronic servicing of mainframe peripheral equipment to include printers and terminals, and experience with installation and maintenance of telecommunication equipment including LAN, fiber optic, and synchronous/asynchronous modems desired. For state application contact NCC Personnel Office, P.O. Box 10, Albany, NY. Deadline for applications is Friday, August 20, EEO/AA Employer.

In another approach, employers desiring to hire new or replacement workers may post job announcements on bulletin boards or other public places. Also, any company's human resources officer or personnel office will usually provide this information on request.

Job information also is available from the local employment service or job service office. These offices, supported by state or local governments, provide assistance in locating jobs.

Any school or college you attend also should provide help in locating job openings. For assistance, contact the school's placement office or career counselor.

Handling Job Interviews

One of the first steps toward landing a job is completing a written job application. If this is required, take these measures:

1. Work carefully in filling out the form.
2. Respond to all questions completely and honestly.
3. Be neat. If possible, use a typewriter or word processor to fill out the application. Otherwise, write neatly and legibly. Always use a pen, not a pencil. Remember that you are trying to make a good impression.
4. Double-check the application. Look for errors in spelling or grammar.
5. If possible, prepare a neatly typed resume in advance. Then attach it to applications or use it instead of application forms, if allowed.

At some point, a completed application will be followed by a job interview. The interview actually is where most jobs are won or lost (the written application serves only to get you to this point). To make the most of any interview, take steps such as these:

1. Always be on time. Being late makes a bad impression. It may cost you the job.
2. Don't appear too eager. A calm, professional manner works best. Act interested, but not desperate.
3. Dress neatly. Always be sure to wear clean, neat clothes.
4. Make efforts at conversation. Even though it is the interviewer's job to ask most of the questions, pose some of your own. But be sure to focus on details about the company or the job itself, not salary or wages. Such details can be discussed in more detail if you are offered the job.

Sooner or later, an interview should lead to a job. Then, it is up to you to make the most of it. With hard work and a positive attitude, a career in installation and repair can be an enjoyable and long-term experience!

BIBLIOGRAPHY

Baxter, Neale. A Positive Connection: Electronics and Careers. *Occupational Outlook Quarterly.* Winter 1989–90, 16–27.

Become a Refrigeration, Heating and Air Conditioning Service Engineer. Refrigeration Service Engineers Society. (Not dated.)

Career Education that Works for America. Career College Association. 1992.

Career Opportunities in the Telephone Industry. United States Telephone Association. (Not dated.)

Fisher, Roger A., and Ken Chernoff. Air Conditioning and Refrigeration Repair. TAB Books. 1988.

Langley, Billy C. Heating, Ventilating, Air Conditioning and Refrigeration. Prentice-Hall. 1990.

Rudman, Jack. Television Servicing. National Learning Corporation. 1991.

Today's Apprentice. Virginia Department of Labor and
    Industry. (Not dated.)
Dictionary of Occupational Titles. VGM Career Horizons.
    1987.
Occupational Outlook Handbook. VGM Career Horizons.
    1992.

SELECTED ORGANIZATIONS RELATED TO REPAIR AND INSTALLATION CAREERS

American Electronics
    Association
    5201 Great American
    Parkway
    Suite 520
    Santa Clara, CA 95054

Associated General
    Contractors of
    America
    1957 E. Street NW
    Washington, DC 20006

Career College Association
    750 First Street, NE
    Washington, DC 20002

Electronic Industries
    Association
    1722 I Street, NW
    Washington, DC 20006

The Electronics
    Technicians
    Association
    602 North Jackson Street
    Greencastle, IN 46135

International Federation of
    Professional and
    Technical Engineers
    8701 Georgia Avenue
    Silver Spring, MD 20910

International Society of
    Certified Electronics
    Technicians
  2708 West Berry Street
  Fort Worth, TX 76109

International Union of
    Electronics,
    Electrical, Salaried,
    Machine and
    Furniture Workers
  1126 16th Street, NW
  Washington, DC 20036

National Automatic
    Merchandising
    Association
  20 North Wacker Drive
  Chicago, IL 60606-3102

Refrigeration Service
    Engineers Society
  International
    Headquarters
  1666 Rand Road
  Des Plaines, IL 60016

United States Telephone
    Association
  900 19th Street, NW,
    Suite 800
  Washington, DC
    20006-2102

SCHOOLS OFFERING TRAINING PROGRAMS

Many private trade and technical schools provide training in electronics, air conditioning and refrigeration technology, computer repair, vending machine repair, or other career areas related to repair or installation. Following are some representative schools accredited by the Career College Association to offer instructional programs in one or more of these fields.

The term *career college* usually does not refer to a college in the traditional sense, but rather to a business school, trade school, or technical school. Programs offered by these schools are designed as an alternative to other types of instruction. A major selling point is that, in many cases, students study only the subjects in which they are interested as opposed to general courses in English, math, social sciences, and so on. At the same time, courses offered by career colleges may not be considered to be college level and, therefore, may not be accepted as transfer credits by colleges or universities. For other advantages and disadvantages, see Chapter 5 on educational options. For a listing of community, junior, and technical colleges, see Appendix C.

For more information, check the yellow pages or consult a high school counselor. Or write to the Career College Association, 750 First Street, NE, Washington, DC 20002.

Alabama

Phillips Junior College
   4900 Corporate Drive
   Huntsville, AL 35805

Alaska

Alaska Junior College
   800 East Diamond
      Boulevard
   Anchorage, AK 99515

Charter College
   2221 East Northern
      Lights Boulevard
   Anchorage, AK 99508

Arizona

North American College
   52108 East Thomas
      Road
   Phoenix, AZ 85016

Arkansas

Southern Technical
      College
   7601 Scott Hamilton
      Drive
   Little Rock, AR 72209

California

Computer Learning
    Center
    3130 Wilshire
    Boulevard
    Los Angeles, CA 90010

Eldorado College
    385 North Escondido
    Boulevard
    Escondido, CA 92025

National Education Center
    (Sawyer Campus)
    5500 South Eastern
    Avenue
    Commerce, CA 90040

Sawyer College at
    Ventura
    470 East Thompson
    Boulevard
    Ventura, CA 93001

Colorado

Pueblo College of
    Business and
    Technology
    330 Lake Avenue
    Pueblo, CO 81004

Connecticut

Computer Processing
    Institute
    305 Boston Avenue
    Stratford, CT 06497

Connecticut Business
    Institute
    605 Broad Street
    Stratford, CT 06497

Data Institute
    745 Burnside Avenue
    East Hartford, CT 06108

District of Columbia

National Education Center
    (Capitol Hill Campus)
    810 First Street NE
    Washington, DC 20002

Florida

Florida Technical College
    8711 Lone Star Road
    Jacksonville, FL 32211

Florida Computer and
    Business School
    8300 West Flagler Street
    Miami, FL 33144

Georgia

Phillips Junior College
    1622 13th Avenue
    Columbus, GA 31901

Illinois

Taylor Business Institute
    36 South State Street
    Chicago, IL 60603

Kentucky

Watterson Career Center
    915 South Third Street
    Louisville, KY 40203

Louisiana

Ayers Institute
    Box 3941
    Shreveport, LA 71105

Southern Technical
    College
    303 Rue Louis XIV
    Lafayette, LA 70508

Phillips Junior College
    822 South Clearview
        Parkway
    New Orleans, LA 70123

Maryland

National Education
    Center, Temple
    School Campus
    3601 O'Donnell Street
    Baltimore, MD 21224

Massachusetts

Computer Learning Center
5 Middlesex Avenue
Somerville, MA 02145

Computer Processing
Institute
615 Massachusetts
Avenue
Cambridge, MA 02139

Salter School
155 Ararat Street
Worcester, MA 01606

Michigan

Great Lakes Junior
College of Business
1231 Cleaver Road
Caro, MI 48723

Mississippi

Phillips Junior College
2680 Insurance Center
Drive
Jackson, MS 39216

Phillips Junior College of
the Mississippi Gulf
Coast
942 Beach Drive
Gulfport, MS 39507

Nevada

Morrison College/Reno
Business College
140 Washington Street
Reno, NV 89503

New Jersey

Cittone Institute
1697 Oak Tree Road
Edison, NJ 08820

Computer Learning Center
160 East Route 4
Paramus, NJ 07652

Dover Business College
15 East Blackwell Street
Dover, NJ 07801

Global Business Institute
33 Journal Square
Jersey City, NJ 07306

New Mexico

Parks College
1023 Tijeras NW
Albuquerque, NM 87102

New York

Bryant and Stratton
Business Institute
1028 Main Street
Buffalo, NY 14202

Bryant and Stratton
Business Institute
792 Calkins Road
Rochester, NY 14623

Grumman Data Systems
Institute
250 Crossways Park
Drive
Woodbury, NY 11797

Pace Business School
45 Park Avenue
Yonkers, NY 10703

SCS Business and
Technical Institute
2467 Jerome Avenue
Bronx, NY 10468

Spencer Business and
Technical Institute
200 State Street
Schenectady, NY 12305

Oklahoma

Draughton College
9301 South Western
Oklahoma City, OK
73501

Oklahoma Junior
College
3232 Northwest 65th
Oklahoma City, OK
73116

Pennsylvania

Berean Institute
　1901 West Girard
　　Avenue
　Philadelphia, PA 19130

Computer Learning Center
　3600 Market Street
　University City, PA
　　19104

National Education
　　Center, Thompson
　　Campus
　5650 Derry Street
　Harrisburg, PA 17111

Pace Institute
　606 Court Street
　Reading, PA 19601

Tennessee

Edmonson Junior College
　　of Business
　3635 Brainerd Road
　Chattanooga, TN 37411

Texas

American Commercial
　　College
　2007 34th Street
　Lubbock, TX 79411

Mansfield Business
　　School
　2800 South IH 35
　Austin, TX 78704

Virginia

National Education
　　Center, Kee Business
　　College Campus
　6301 Midlothian
　　Turnpike
　Richmond, VA 23225

Commonwealth College
　4160 Virginia Beach
　　Boulevard
　Virginia Beach, VA
　　23452

Virginia College
  4142 Melrose Avenue
    NW
  Roanoke, VA 24017

Wisconsin

Stratton College
  1300 North Jackson
    Street
  Milwaukee, WI 53202

TWO-YEAR COLLEGES OFFERING PROGRAMS IN REPAIR AND INSTALLATION FIELDS

Many two-year colleges offering programs in computer technology, electronics, or other fields involving installation or repair. Located throughout the United States and Canada, over 1,200 such institutions serve their local areas.

For most community, technical, or junior colleges, students commute to classes rather than live on campus. Many of these colleges offer certificate or degree programs in technical areas related to repair or installation of equipment or other items. For any college in which you are interested, check with the school's admissions office to find out what programs are available.

The following is a partial list of colleges offering programs related to installation and repair. For information about a school near you or any other two-year college not listed, consult the school's catalog for more information.

Alabama

John C. Calhoun State
    Community College
    Decatur, AL 35602

Gadsen State Community
    College
    Gadsen, AL 35999

Jefferson State
    Community College
    Birmingham, AL 35215

George C. Wallace State
    Community College
    Dothan, AL 36303

Arizona

Arizona Western College
    Yuma, AZ 85364

Cochise College
    Douglas, AZ 85607

Eastern Arizona College
    Thatcher, AZ 85552

Glendale Community
    College
    Glendale, AZ 85302

Northland Pioneer College
    Holbrook, AZ 86025

Phoenix College
    Phoenix, AZ 85013

Pima Community College
    Tucson, AZ 85709

Rio Salado Community
    College
    Phoenix, AZ 85003

Yavapai College
    Prescott, AZ 86301

Arkansas

Arkansas State
    University—Beebe
    Branch
    Beebe, AR 72012

East Arkansas Community
    College
    Forrest City, AR 72335

Garland County
    Community College
    Hot Springs, AZ 71913

Mississippi County
    Community College
    Blytheville, AR 72316

Phillips County
    Community College
    Helena, AR 72342

Westark Community
    College
    Forth Smith, AR 72913

California

Cabrillo College
    Aptos, CA 95003

Chabot College
    Hayward, CA 94545

Chaffey Community
    College
    Rancho Cucamonga, CA
    91701

College of the Redwoods
    Eureka, CA 95501

Cuyamaca College
    El Cajon, CA 92019

Cosumnes River College
    Sacramento, CA 95823

Fullerton College
    Fullerton, CA 92634

Golden West College
    Huntington Beach, CA
    92647

Los Angeles City College
    Los Angeles, CA 90029

Los Angeles Pierce
    College
    Woodland Hills, CA
    91371

Mission College
Santa Clara, CA 95054

Mt. San Antonio College
Walnut, CA 91789

Mt. San Jacinto College
San Jacinto, CA 92383

Napa Valley College
Napa, CA 94558

Orange Coast College
Costa Mesa, CA 92626

Rancho Santiago
Community College
Santa Ana, CA 92706

Santa Rosa Junior
College
Santa Rosa, CA 95401

Yuba College
Marysville, CA 95901

Colorado

Aims Community
College
Greeley, CO 80631

Community College of
Aurora
Aurora, CO 80011

Front Range Community
College
Westminster, CO 80030

Morgan Community
College
Ft. Morgan, CO 80701

Northeastern Junior
College
Sterling, CO 80751

Pikes Peak Community
College
Colorado Springs, CO
80906

Pueblo Community
    College
Pueblo, CO 81004

Red Rocks Community
    College
Lakewood, CO 80401

Connecticut

Norwalk State Technical
    College
Norwalk, CT 06854

Delaware

Delaware Technical and
    Community College
Neward, DE 19702

Florida

Brevard Community
    College
Cocoa, FL 32922

Indian River Community
    College
Fort Pierce, FL 34981

Lake City Community
    College
Lake City, FL 32055

Manatee Community
    College
Bradenton, FL 32406

Miami-Dade Community
    College
Miami, FL 33132

Palm Beach Community
    College
Lake Worth, FL 33461

Santa Fe Community
    College
Gainesville, FL 32601

Seminole Community
    College
Sanford, FL 32771

Georgia

Brunswick Junior College
  Brunswick, GA 31523

Gainesville College
  Gainesville, GA 30503

Middle Georgia College
  Cochran, GA 31014

Waycross Junior College
  Waycross, GA 31501

Hawaii

University of
  Hawaii-Honolulu
  Community College
  Honolulu, HI 96817

University of
  Hawaii-Kauai
  Community College
  Lihue, HI 96766

University of Hawaii-Hilo
  Community College
  Hilo, HI 96720

Idaho

Eastern Idaho Technical
  College
  Idaho Falls, ID 83404

North Idaho College
  Coeur D'Alene, ID
  83814

Ricks College
  Rexburg, ID 83440

Illinois

Belleville Area College
  Belleville, IL 62221

Black Hawk College
  Moline, IL 61265

College of Du Page
  Glen Ellyn, IL 60137

Elgin Community
    College
Elgin, IL 60120

John A. Logan College
    Carterville, IL 62918

Joliet Junior College
    Joliet, IL 60436

Highland Community
    College
Freeport, IL 61032

Kennedy-King College
    Chicago, IL 60621

Lewis and Clark
    Community College
Godfrey, IL 62035

Kishwaukee College
    Malta, IL 60150

McHenry County College
    Crystal Lake, IL 60012

Moraine Valley
    Community College
Palos Hills, IL 60465

Richard J. Daley College
    Chicago, IL 60652

Rock Valley College
    Rockford, IL 61101

Shawnee Community
    College
Ullin, IL 62992

Triton College
    River Grove, IL 60171

Southeastern Illinois
    College
Harrisburg, IL 62946

Wabash Valley College
    Mt. Carmel, IL 62863

William Rainey Harper
    College
Palatine, IL 60067

Indiana

Indiana Vocational
Technical College—
Central Indiana
Indianapolis, IN 46202

Indiana Vocational
Technical College—
Northcentral
South Bend, IN 46619

Indiana Vocational
Technical College—
Southwest
Evansville, IN 47710

Iowa

Clinton Community
College
Clinton, IA 52732

Iowa Central Community
College
Fort Dodge, IA 50501

Iowa Western Community
College
Council Bluffs, IA 51501

Kirkwood Community
College
Cedar Rapids, IA 52406

Northwest Iowa Technical
College
Sheldon, IA 51201

Western Iowa Technical
Community College
Sioux City, IA 51102

Kansas

Allen County Community
College
Iola, KS 66749

Cowley County
Community College
Arkansas City, KS 67005

Butler County Community
    College
El Dorado, KS 67042

Garden City Community
    College
Garden City, KS 67846

Haskell Indian Junior
    College
Lawrence, KS 66044

Johnson County
    Community College
Overland Park, KS
    66210

Kansas City Kansas
    Community College
Kansas City, KS 66112

Neosho County
    Community College
Chanute, KS 66720

Kentucky

Maysville Community
    College
Maysville, KY 41956

Owensboro Community
    College
Owensboro, KY 42301

Maryland

Catonsville Community
    College
Baltimore, MD 21228

Dundalk Community
    College
Dundalk, MD 21222

Frederick Community
    College
Frederick, MD 21701

Massachusetts

Massasoit Community
    College
    Brockton, MA 02402

Middlesex Community
    College
    Bedford, MA 01730

Springfield Technical
    Community College
    Springfield, MA 01105

Michigan

Alpena Community
    College
    Alpena, MI 49707

Delta College
    University Center, MI
    48710

Henry Ford Community
    College
    Dearborn, MI 48128

Gogebic Community
    College
    Ironwood, MI 49938

Kellogg Community
    College
    Battle Creek, MI 49016

Macomb County
    Community College
    Warren, MI 48093

Northwestern Michigan
    College
    Traverse City, MI 49684

Oakland Community
    College
    Bloomfield Hills, MI
    48013

Schoolcraft College
    Livonia, MI 48152

West Shore Community
    College
    Scottville, MI 49454

Minnesota

Anoka-Ramsey
   Community College
   Coon Rapids, MN 55433

Lakewood Community
   College
   White Bear Lake, MN
   55110

Northeast Metro Technical
   Institute
   White Bear Lake, MN
   55110

North Hennepin
   Community College
   Brooklyn Park, MN
   55445

Wilmar Technical College
   Wilmar, MN 56201

Mississippi

East Central Junior College
   Decatur, MS 39327

Hinds Community
   College
   Raymond, MS 39514

Holmes Junior College
   Goodman, MS 39079

Itawamba Community
   College
   Fulton, MS 38843

Jones County Junior
   College
   Ellisville, MS 39437

Mississippi Gulf
   Coast Community
   College
   Perkinston, MS 39573

Mississippi Delta Junior
   College
   Moorhead, MS 38761

Pearl River Community
   College
   Poplarville, MS 39470

Missouri

East Central College
Union, MO 63084

Maple Woods Community
College
Kansas City, MO 64156

Moberly Area Junior
College
Moberly, MO 65270

Penn Valley Community
College
Kansas City, MO 64131

St. Louis Community
College
St. Louis, MO 63102

Nebraska

Central Community
College, Platte
Campus
Columbus, NE 68601

Northeast Community
College
Norfolk, NE 68701

Mid-Plains Community
College
North Platte, NE 69101

Western Nebraska
Community College
Sidney, NE 69162

Nevada

Northern Nevada
Community College
Elko, NV 89801

Truckee Meadows
Community College
Reno, NV 89512

Western Nevada
Community College
Carson City, NV 89701

New Hampshire

New Hampshire Technical
    Institute
Concord, NH 03302

New Jersey

Atlantic Community
    College
Mays Landing, NJ 08330

Burlington County College
Pemberton, NJ 08068

Mercer County
    Community College
Trenton, NJ 08690

Ocean County College
Toms River, NJ 08754

New Mexico

Eastern New Mexico
    University—Clovis
Clovis, NM 88101

Eastern New Mexico
    University—Roswell
Roswell, NM 88202

Northern New Mexico
    Community College
El Rito, NM 87530

New Mexico State
    University—Carlsbad
Carlsbad, NM 88220

New York

Adirondack Community
    College
Queensburg, NY 12804

Corning Community
    College
Corning, NY 14830

Orange County
    Community College
Middletown, NY 10940

Suffolk County
    Community College
Selden, NY 11784

North Carolina

Ashville Buncombe
    Technical College
Ashville, NC 28801

Cape Fear Community
    College
Wilmington, NC 28401

Catawba Valley
    Community College
Hickory, NC 28601

Central Piedmont
    Community College
Charlotte, NC 28235

Coastal Carolina
    Community College
Jacksonville, NC 28540

College of the
    Albermarle
Elizabeth City, NC
    27906

Forsyth Technical
    Community College
Winston-Salem, NC
    27103

Isothermal Community
    College
Spindale, NC 28160

Johnston Community
    College
Smithfield, NC 22577

Lenoir Community
    College
Kinston, NC 28501

McDowell Technical
    Community College
Marion, NC 28752

Robeson Community College
Lumberton, NC 28359

Surry Community College
Dobson, NC 27017

Wake Technical College
Raleigh, NC 27603

Wilkes Community College
Wilkesboro, NC 28697

Ohio

Cuyahoga Community College
Cleveland, OH 44115

Columbus State Community College
Columbus, OH 43216

Edison State Community College
Piqua, OH 45356

Lakeland Community College
Mentor, OH 44060

Stark Technical College
Canton, OH 44720

Washington Technical College
Marietta, OH 45750

Oklahoma

El Reno Junior College
El Reno, OK 73036

Northeastern Oklahoma Agricultural and Mechanical College
Miami, OK 74354

Rose State College
Midwest City, OK 73110

Tulsa Junior College
Tulsa, OK 74119

Oregon

Central Oregon
Community College
Bend, OR 97701

Chemeketa Community
College
Salem, OR 97309

Clatsop Community
College
Astoria, OR 97103

Linn-Benton Community
College
Albany, OR 97321

Mount Hood Community
College
Gresham, OR 97030

Portland Community
College
Portland, OR 97219

Rogue Community
College
Grants Pass, OR 97526

Southwestern Oregon
Community College
Coos Bay, OR 97420

Treasure Valley
Community College
Ontario, OR 97914

Pennsylvania

Butler County Community
College
Butler, PA 16001

Community College of
Allegheny
County—Boyce
Monroeville, PA 15146

Community College of
  Allegheny
  County—North
Pittsburgh, PA 15237

Community College of
  Allegheny
  County—South
West Mifflin, PA 15122

Community College of
  Beaver County
Monaca, PA 15061

Harrisburg Area
  Community College
Harrisburg, PA 17110

Lehigh County
  Community College
Schnecksville, PA 18078

Pennsylvania College of
  Technology
Williamsport, PA 17701

Westmoreland County
  Community College
Youngwood, PA 15697

South Carolina

Aiken Technical College
Aiken, SC 29801

Denmark Technical
  College
Denmark, SC 29042

Greenville Technical
  College
Greenville, SC 29606

Florence Darlington
  Technical College
Florence, SC 29501

Horry-Georgetown
  Technical College
Conway, SC 29526

Piedmont Technical
    College
  Greenwood, SC 29646

Spartanburg Technical
    College
  Spartanburg, SC 29303

Sumter Area Technical
    College
  Sumter, SC 29150

Trident Technical College
  Charleston, SC 29411

Tri-County Technical
    College
  Pendleton, SC 29670

Tennessee

Chattanooga State
    Technical Community
    College
  Chattanooga, TN 37406

Cleveland State
    Community College
  Cleveland, TN 37320

Jackson State Community
    College
  Jackson, TN 38301

Nashville State Technical
    Institute
  Nashville, TN 37209

Northeast State Technical
    Community College
  Blountville, TN 37617

State Technical Institute at
    Memphis
  Memphis, TN 38134

Tri-Cities State Technical
    Institute
  Blountville, TN 37617

Walters State Community
    College
  Morristown, TN 37813

Texas

Angelina College
 Lufkin, TX 75902

Central Texas College
 Killeen, TX 76541

El Paso Community
 College
 El Paso, TX 79998

Grayson County College
 Denison, TX 75020

Houston Community
 College System
 Houston, TX 77270

Kilgore College
 Kilgore, TX 75662

Laredo Junior College
 Laredo, TX 78040

Midland College
 Midland, TX 79701

Odessa College
 Odessa, TX 79762

Richland College
 Dallas, TX 75243

San Antonio College
 San Antonio, TX 78212

South Plains College
 Levelland, TX 79336

Texarcana College
 Texarcana, TX 75501

Tarrant County Junior
 College
 Forth Worth, TX 76102

Texas Southmost College
 Brownsville, TX 78520

Utah

Salt Lake Community
 College
 Salt Lake City, UT
 84130

Utah Valley Community
   College
   Orem, UT 84058

Virginia

Dabney S. Lancaster
   Community College
   Clifton Forge, VA 24422

Central Virginia
   Community College
   Lynchburg, VA 24502

John Tyler Community
   College
   Chester, VA 23831

Lord Fairfax Community
   College
   Middletown, VA 22645

Mountain Empire
   Community College
   Big Stone Gap, VA
   24219

New River Community
   College
   Dublin, VA 24084

Northern Virginia
   Community College
   Annandale, VA 22003

Patrick Henry Community
   College
   Martinsville, VA 24112

Southside Virginia
   Community College
   Alberta, VA 23821

Southwest Virginia
   Community College
   Richlands, VA 24641

Thomas Nelson
   Community College
   Hampton, VA 23670

Tidewater Community
   College
   Portsmouth, VA 23703

Virginia Western
    Community College
    Roanoke, VA 24015

Wytheville Community
    College
    Wytheville, VA 24382

Washington

Centralia College
    Centralia, WA 98531

Clark Community College
    Vancouver, WA 98663

Columbia Basin
    Community College
    Pasco, WA 99301

Edmonds Community
    College
    Lynnwood, WA 98036

Pierce College
    Tacoma, WA 98498

Skagit Valley College
    Mount Vernon, WA
    98273

Spokane Community
    College
    Spokane, WA 99207

Walla Walla Community
    College
    Walla Walla, WA 99362

West Virginia

West Virginia Northern
    Community College
    Wheeling, WV 26003

West Virginia University
    at Parkersburg
    Parkersburg, WV 26101

Wisconsin

Black Hawk Technical
    College
    Janesville, WI 53547

Gateway Technical College
  Kenosha, WI 53141

Madison Area Technical
  College
  Madison, WI 53703

Northern Wisconsin
  Technical College
  Green Bay, WI 54307

Wyoming

Casper College
  Casper, WY 82601

Central Wyoming College
  Riverton, WY 82501

Canadian Institutions

For information about programs offered by Canadian two-year colleges, contact the institution nearest you. Or write to

American Association of Community Colleges
  One Dupont Circle, NW
  Washington, DC 20036.

About twenty Canadian colleges are members of this organization.

FIELD AND REGIONAL OFFICES OF THE BUREAU OF APPRENTICESHIP TRAINING, U.S. DEPARTMENT OF LABOR

REGION I

CONNECTICUT

Hartford

Room 367
Federal Building
135 High Street, 06103

MAINE

Augusta

Room 408-D
Federal Building
68 Sewall Street, 04330

MASSACHUSETTS

Boston

11th Floor
One Congress Street, 02114

Springfield	Room 211
	Springfield Federal Building
	1550 Main Street, 01103
Worcester	Room 316, Federal Building
	595 Main Street, 01601

NEW HAMPSHIRE

| Concord | 143 North Main Street, 03301 |

RHODE ISLAND

| Providence | Providence Federal Building |
| | 100 Hartford Avenue, 02909 |

VERMONT

Burlington	Suite 103
	Burlington Square
	96 College Street, 05401

REGION II

NEW JERSEY

| Iselin | Parkway Towers |
| | Building E, Third Floor, 08830 |

NEW YORK

Albany	Leo O'Brien Federal Building North Pearl and Clinton Avenue, 12202
New York	Room 602 Federal Building 201 Varick Street, 10014
Buffalo	Room 209, Federal Building 111 West Huron Street, 14202
Rochester	100 State Street Federal Building, 14614
Syracuse	Federal Building 100 South Clinton Street, 13260

REGION III

DELAWARE

Wilmington	Federal Building 844 King Street, 19801

DISTRICT OF COLUMBIA

D.C.I.	Room 517 1111-20th Street, NW, 20036

MARYLAND

Baltimore	Room 1028 31 Hopkins Plaza Charles Center, 21201

PENNSYLVANIA

Erie	Room 106, Federal Building 6th and State Streets, 16507
Harrisburg	Federal Building 228 Walnut Street, 17108
Philadelphia	Room 13240 3535 Market Street, 19104
Pittsburgh	Room 1436 Federal Building 1000 Liberty Avenue, 15222
Reading	Room 2115, East Shore Office Building 45 South Front Street, 19603
Wilkes-Barre	Room 2028, Penn Place 20 North Pennsylvania Avenue, 18701

VIRGINIA

Richmond	Room 10-020 400 North 8th Street, 23240

WEST VIRGINIA

Clarksburg   Room 701
                  Palace Furniture Building
                  168 West Main Street, 26301

REGION IV

ALABAMA

Birmingham   Suite 102
                  Berry Building
                  2017 2nd Avenue, North, 35203

FLORIDA

Miami   995 North West 119th Street, 33168

Tallahassee   Suite 264
                  2574 Seagate Drive, 32301

GEORGIA

Atlanta   Suite 200
                  1371 Peachtree Street, NE, 30367

Columbus   Suite 24
                  3604 Macon Road, 31907

Savannah   Suite 303
                  120 Barnard Street, 31801

KENTUCKY

Louisville   Room 187-J
                    Federal Building
                    600 Federal Place, 40202

Lexington   2033 Regency Road, 40503

MISSISSIPPI

Jackson   Suite 1010
                    Federal Building
                    101 West Capitol Street, 39269

NORTH CAROLINA

Raleigh   Suite 375
                    Somerset Center
                    4505 Falls of the Neuse Road, 27609

SOUTH CAROLINA

Columbia   Suite 838
                    Strom Thurmond Federal Building
                    1835 Assembly Street, 29201

Charleston   Suite 313, Federal Building
                    334 Meeting Street, 29403

TENNESSEE

Nashville   Suite 101-A
                    460 Metroplex Drive, 37211

REGION V

ILLINOIS

Chicago

Room 758
230 South Dearborn Street, 60604

Alton

501 Belle Street, 62002

Des Plaines

1420 Miner Street, 60016

Peoria

100 North East Monroe Street, 61602

Rockford

211 South Court Street, 61101

Springfield

U.S. Post Office, Room 14
600 East Monroe Street, 62701

INDIANA

Indianapolis

Room 414
Federal Building
46 East Ohio Street, 46204

MICHIGAN

Detroit

Room 657
231 West Lafayette Avenue, 48226

MINNESOTA

St. Paul

Room 134
Federal Building
316 Robert Street, 55101

Duluth	Room 234 515 West First Street, 55802

OHIO

Cleveland	Suite 602 1375 Euclid Avenue, 44115

WISCONSIN

Madison	Room 303, Federal Center 212 East Washington Avenue, 53703

REGION VI

ARKANSAS

Little Rock	Room 3507 Federal Building 700 West Capitol Street, 72201

LOUISIANA

New Orleans	Room 1323, U.S. Postal Building 701 Loyola Street, 70115

NEW MEXICO

Albuquerque	Room 830 505 Marquette, 87102

OKLAHOMA

Tulsa   Room 305, 51 Yale Building
                   5110 South Yale, 74135

TEXAS

Dallas   Room 502, Federal Building
                   525 Griffin Street, 75202

REGION VII

IOWA

Des Moines   Federal Office Building
                   210 Walnut Street, 50309

KANSAS

Kansas City   Room 1100 Federal Office Building
                   911 Walnut Street, 64016

Wichita   Room B-41
                   U.S. Courthouse Building
                   401 North Market, 67202

MISSOURI

St. Louis   Robert A. Young Federal Building
                   1222 Spruce Street, 63103

NEBRASKA

Omaha

Room 801
106 South 15th Street, 68102

REGION VIII

COLORADO

Denver

Room 476
U.S. Custom House
721 19th Street, 80202

MONTANA

Helena

Room 394
Federal Office Building
301 South Park Avenue, 59626

NORTH DAKOTA

Fargo

Room 428
New Federal Building
653 2nd Avenue N, 58102

SOUTH DAKOTA

Sioux Falls

Room 107
Courthouse Plaza
300 North Dakota Avenue, 57102

UTAH

Salt Lake City

Room 1051
Administration Building
1745 West 1700 South, 84101

WYOMING

Cheyenne

Room 5013
Joseph C. Mahoney Federal Center
2120 Capitol Avenue, 82201

REGION IX

ARIZONA

Phoenix

3221 North 16th Street, 85016

CALIFORNIA

San Francisco

Suite 715
71 Stevenson Street, 94105

HAWAII

Honolulu

Room 5113
300 Ala Moana Boulevard, 96580

NEVADA

Las Vegas

Room 311, U.S. Court House
301 Stewart Avenue, 89101

REGION X

ALASKA

Anchorage   Room 554, Federal Building
222 West 7th Street, 99513

IDAHO

Boise   Suite 128
3050 North Lakeharbor Lane, 83703

OREGON

Portland   526 Federal Building
1220 South West 3rd Avenue, 97204

WASHINGTON

Seattle   Room 925
1111 Third Avenue, 98101